The Fast Track to (Finally!) Getting on the Air With Ham Radio

Michael Burnette, AF7KB with Kerry Burnette, KC7YL

The Fast Track to (Finally!) Getting on the Air With Ham Radio

All content © Copyright 2020 Michael Burnette. All rights reserved. This book or any portion thereof may not be reproduced or used in any manner whatsoever without the express written permission of the publisher except for the use of brief quotations in a book review. All trademarks, service marks, trade names, trade dress, product names and logos appearing in this work are the property of their respective owners. Any rights not expressly granted herein are reserved.

The author assumes no liability for your use of the instructional materials provided in this book. The author is neither a professional contractor, mobile installer, nor electrician. Your safety is your responsibility.

Dedication

For Delvin Bunton, NS7U

Thanks for all you've done for us, for ham radio, and for your community, Delvin.

Table of Contents

	Dedication	iii
I	**What's Stopping You?**	**1**
1	**Introduction**	**3**
	"I Have no Idea How to Build a Station"	6
	"There Are So Many Rules! I'm Sure to Break One!"	7
	"I'm a Little …uh … Mic Shy…"	15
II	**Getting Started on the VHF/UHF Bands**	**19**
2	**Working Repeaters**	**21**
	Finding Your Local Repeaters	22
3	**Dual-band Handheld Transceivers**	**25**
	Manually Programming a Baofeng UV-5R	28
	Programming a UV-5R With a Computer	31
	Antennas for Handhelds	37
4	**Dual-band Mobile Station**	**41**
	Dual-band Mobile Transceivers	42
	Installing Your Dual-band Mobile Radio	45
	Professional Installation	47
	Doing it Yourself	48
	All the Parts	51
	Tools	58
	Let's Install a Radio!	62

	Plan, Plan, Plan, then Plan Some More	63
	Power	64
	Antenna	66
	Feed Line	67
	Mounting the Radio/Control Head	69
	Final Steps	70
5	**Dual-band Fixed Stations**	**77**
	Choosing a Fixed Station Radio	77
	Your VHF/UHF "Shack"	78
	Power Supply	78
	Antenna	81
	Antenna Mounts	83
	Feed Line	84
6	**Grounding the Amateur Radio Station**	**91**
	Fundamentals of Grounding	93
	Ground Rods	96
	Safety Grounding	97
	RF Grounding	99
	Fighting Common Mode Currents	101
	Lightning Protection Grounds	104
	Lightning Basics	105
	Ground Loops	110
7	**Operating on VHF/UHF**	**113**
	Dual-band Radio Controls	113
	Repeaters	114
	Nets	118
	EchoLink®	121
8	**Advanced VHF/UHF**	**125**
	Simplex Operation	125
	Contesting	128
	Digital Voice Modes	131
	Digital Hot Spots	134

III Getting Started in HF — 137

9 Building Your First HF Station — 139
- Planning Your HF Station 142
  - HF Antennas 144
  - HF Feed Lines 154
- Choosing Your First HF Transceiver 156
  - Power Supply 160
  - Antenna Tuner 161

10 HF Transceiver Controls — 163

11 Operating on HF — 171
- CW 171
- SSB Phone 172
- HF Nets 174
- HF Contesting 176
- Search-and-pounce on HF 177
- DX on HF 177
- Digital Modes on HF 183

12 Closing Thoughts — 189

Amateur Radio Glossary — 191

About the Authors — 227

List of Tables

1.1	Part 97 Musts and Must Nots	12
1.2	"I should look this up..."	13
1.3	RF Exposure Evaluation Requirements	14
3.1	How to Program a Baofeng UV-5R	31
4.1	Antenna Tuning With an SWR Meter	75
5.1	Coaxial Cable Losses	87
7.1	NATO Phonetic Alphabet	115
8.1	2-meter and 70-cm Simplex Frequencies	125
8.2	June VHF Contest Rules	129
11.1	RST Signal Reporting System	175
11.2	Common Phone Q Codes	176
12.1	Amateur Radio Glossary	219

List of Figures

1.1	RF Exposure Calculator Data Entry Page	10
1.2	RF Exposure Calculator Results Page	11
2.1	RepeaterBook Listing for WA7LAW	23
3.1	Yaesu FT3DR Handheld Transceiver	25
3.2	Tram Amateur Dual-Band Base Antenna	26
3.3	Baofeng UV-5R	27
3.4	Baofeng Programming Cable	32
3.5	CHIRP Programming Screen	33
3.6	Yaesu FT-2DR	36
3.7	Kenwood TH-D72A	36
3.8	Yaesu FT-65R	37
3.9	Dual-band J-Pole Antenna (Courtesy of "J-Pole John" Bole, K9JEB https://k9jeb.com/JPole.aspx)	38
3.10	Roll-up Slim-Jim Antenna Kit by The Survival Antenna	39
3.11	Buddipole Portable Multi-band Antenna	40
4.1	Yaesu FTM-400DR Mobile Transceiver	42
4.2	Kenwood TM-D710GA	43
4.3	AF7KB & KC7YL's Kenwood Install	44
4.4	Wouxun KG-UV920P	44
4.5	QYT KT-980 (left) & BTECH UV-50X2 (right)	45
4.6	Police Car Interior	49
4.7	Magnetic Mount Antenna	52
4.8	Comet CTC-50M "Window Gap Jumper"	53
4.9	The Famous Little Red Porcupine Car	54
4.10	NMO Through-hole Mount & Connector	55

4.11	Comet Fender-style Antenna Mount	55
4.12	MFJ-348M Trunk Lip Mount	56
4.13	Thin Feed Line for Door Pass-through	57
4.14	InstallGear Battery Terminals	57
4.15	Water-resistant Butt Connector	58
4.16	Pliers: Slip-joint, Diagonal, Lineman's, Long-nose	59
4.17	Socket Set	60
4.18	Crimping Tool	60
4.19	Wire Strippers	60
4.20	Trim & Molding Tool Set	61
4.21	Multimeter	62
4.22	Continuity Tester	62
4.23	Wire Insertion Tool	65
4.24	Wire Loom and Fuse	65
4.25	Antenex NMO Installation Saw	67
4.26	Flush Cutters	69
4.27	Lido Gooseneck Seat Bolt Mount	70
4.28	Lido Cupholder Mount	71
4.29	Checking for Feed Line Short Circuit	71
4.30	MFJ-269C Antenna Analyzer	72
4.31	(Ideal) SWR Curve	72
4.32	SWR Meter/Wattmeter	73
4.33	Coax Jumper	74
4.34	Antenna Length Adjustment Set Screws	74
5.1	ICOM IC-271A	77
5.2	MFJ-4115 15 Amp Power Supply	79
5.3	Anderson Powerpoles	80
5.4	Cushcraft A270-6S Dual-band Yagi	81
5.5	Gable Mount	83
5.6	Rohn Non-penetrating Flat Roof Mount	84
5.7	Anatomy of a Coaxial Cable	85
5.8	Lightning Surge Protector - Note ground connection at bottom.	90
5.9	Lightning Surge Protector Connection	90
6.1	Equivalent Circuit of Ground	94
6.2	Equivalent Circuit of Effective Ground System	95

6.3	Jumpers with Alligator Clips	95
6.4	Fence Post Driver	97
6.5	Driving Ground Rod with Power Hammer	98
6.6	Schematic of Ground System	98
6.7	One-inch Grounding Braid	99
6.8	Ground Bus Bar	100
6.9	Differential-Mode Current	101
6.10	Common-Mode Current	102
6.11	Snap-on Ferrite Choke	103
6.12	Ferrite Ring	103
6.13	You Don't Want This in Your House	104
6.14	US Airman Installing Lightning Protection - Note size of bonding wire.	107
6.15	An "Almost Right" Ground System - Don't Do This!	108
6.16	Proper Ground System - Do This!	109
6.17	Ground Rod with Bonding Clamp	110
6.18	Split Bolt	111
6.19	How to Create a Ground Loop	111
6.20	No Ground Loop	112
7.1	EchoLink® Home Screen	122
7.2	EchoLink® QSO Screen	123
7.3	EchoLink® Desktop Program	124
8.1	Downtown Seattle, WA	126
8.2	Tape Measure Yagi - Courtesy of Michael Hill, WA7MPH	128
8.3	Yaesu FTM-100DR	132
8.4	Kenwood TH-D74 D-Star Radio	133
8.5	AnyTone AT-D868UV (Photo courtesy of Bridgecomsystems.com)	133
8.6	OpenSpot 2 Digital Hot Spot	134
9.1	This Could be You!	139
9.2	Propagation Forecast	141
9.3	DXMaps.com QSO Map	142
9.4	A Lot of Money in the Form of Bicycles and Gear	143
9.5	Flat-top Antenna	144
9.6	Sloper Antenna	145
9.7	Inverted V Antenna	145

9.8	Ready-made G5RV Antenna	146
9.9	Horizontal Loop Antenna	147
9.10	Hanging a Long-Wire Antenna	148
9.11	Air Boss Antenna Launcher. Photo courtesy of Bill Olah, KR4LO. Used by permission.	148
9.12	43-foot Zerofive Vertical Antenna (Picture courtesy of Floyd Larck, KK3Q)	150
9.13	HyGain DX-88	151
9.14	Radial Plate	151
9.15	Zerofive Flagpole Antenna	152
9.16	Magnetic Loop Antenna	153
9.17	Ham Radio Deluxe Remote Control Screen	155
9.18	MFJ Coax Window Feed-Through	156
9.19	ICOM IC-7851	156
9.20	Collins KWM-2	157
9.21	Alinco DX-SR8T	157
9.22	Yaesu FT-891	158
9.23	ICOM IC-718	158
9.24	ICOM IC-7300	159
9.25	Used Kenwood TS-520S	159
9.26	Astron RS-35A	160
9.27	MFJ-4125 Switch-mode Power Supply	161
9.28	MFJ-949E Versa Tuner II	162
9.29	MFJ-945E	162
10.1	Heathkit DX–40 Transmitter (L) and Heathkit HR-10 Receiver (R) DX–40 Photo Courtesy of AB2RA, Janis Carson. HR-10 Photo Courtesy of K6JCA, Jeff Anderson.	163
10.2	Kenwood TS-530s	164
10.3	Interfering Signal Overlapping Desired Signal	166
10.4	Normal SSB Band-Pass Filtering	166
10.5	Narrower Band-Pass Filtering	166
10.6	IF-Shift	167
11.1	QSL Card	178
11.2	DX Lab SuiteDXView	179
11.3	DXView World Map	180

11.4	DXLab Suite PropView Window	181
11.5	PropView Propagation Forecast	182
11.6	"Textbook" Setup for HF Digital	184
11.7	Real World Setup for HF Digital	184
11.8	Ham Radio Deluxe PSK31 Screen	186
11.9	WSJT-X FT8 Screen	188
11.10	Time.is Screenshot	188

Part I

What's Stopping You?

Chapter 1

Introduction

Here's a little secret about ham radio; *most* of the people who walk out of a Technician Class license exam proudly clutching their CSCE (Certificate of Successful Completion of Examination) have just done the next-to-last thing they'll ever do in ham radio. Their final ham radio action will be checking the FCC web site to see their new call sign. After that, nothing.

It's even worse.

I know at least two people who have had their Extra Class tickets for over a year and *still* have never transmitted a single signal.

How can this possibly be? These folks were motivated enough to at least memorize 74% of the correct answers, then find their way to a VE testing session, pay their fee, and sit for the exam. Then what happened?

I can imagine a few scenarios. Others I don't need to imagine because I've heard them directly from people.

There are the people who honestly never wanted to get involved in the hobby. These might include the spouses/children/friends of enthusiastic hams. My own wife got her Technician license mostly because she was curious about "that ham radio thing you do." She really had very little intention of ever operating a ham radio. In my mind, her obtaining a ham license was such an unlikely event that when she announced, "I think I'll get my license," I actually asked "Uh...what license?" (Then she got hooked – but that's another story.) Other spouses/children/friends we've known took the exam in the hope they just wouldn't have to hear about it any more. You can imagine their level of enthusiasm for the hobby. Without some sort of "I've seen the light!" experience, I really doubt we'll get many of those people to become active hams.

Another group is the preppers, the folks who are busy preparing for various Doomsday scenarios. Lots of prepper web sites urge them to get their ham radio license for emergency communications in that dire time When the Stuff Hits the Fan. I suspect more than a few of them get their license and, figuratively, put it in the basement next to the dried beans and bottled water, feeling – completely mistakenly – that, communications-wise, they're now ready for the worst. While I know some very competent prepper hams, those prepper web sites seldom seem to mention that effective communications in an emergency are totally dependent on *lots* of planning, practice, and teamwork – not to mention, usually, much more equipment than a $25 amazon.com radio.

Fair enough; those are a couple of groups who took that exam for reasons that really had nothing to do with becoming an active amateur radio operator.

That still leaves a lot of people who did have every intention of participating in the hobby but ran out of gas shortly after attaining that license.

What happened to those folks?

Based on a lot of conversations with those people, it often seems to boil down to two factors.

First, there's the result of following this common advice:

> "Just go on the internet and do practice exams until you can pass the test." – 10,000 Hams on Facebook

We've had countless conversations like the following one at various hamfests. It begins with someone sort of staring at our General book, and admitting, "Yeah – I did the Tech exam about a year ago – but I don't really remember any of it."

"Oh, really? How did you prepare for the exam?"

"I just went on line and memorized the answers."

"Ah …"[Awkward pause …]

I realize the advice to "just memorize the answers" comes from the best intentions, but we're not doing anyone any favors by advocating that supposedly easy path. The exam material is slim enough preparation for the hobby as it is, even if they take a comprehensive program like *The Fast Track*, or the ARRL's, or Gordon West's, for that matter. Just memorizing answers is a shortcut to nowhere. I'd suggest it would be best for us to be honest with the people we're recruiting and let them know it takes a bit of knowledge to make ham radio stuff work. It is not "plug'n'play," and if we don't disclose that up front, we're just setting them up for later disappointment.

Even with good, solid prep, though, there's a considerable distance for the beginner to travel from the rules and theory covered on the exams and real, "now what do I do?" practical application. This book is aimed at bridging that gap.

I notice a lot of new hams turn to the resources that people often use these days when they want information or opinions; Facebook's ham radio related groups. As you may have already discovered, those are a deep pool of negativity and misinformation; they are often neither useful nor encouraging to the new ham. You may find me on there from time to time, but I openly admit my Facebook time is strictly for business.

Then there's the anxiety around actually pushing that PTT (Push To Talk) button and, gulp, saying stuff out loud on the radio where everybody can hear it. That's natural enough, and the tests don't help much in that area either, because they leave the impression that talking on the radio is really complicated and requires all sorts of cryptic language. Heaven help the newbie ham who said CQ ("calling anyone") when they should have said QSL ("I acknowledge receipt of the message")! We'll work on that, too, and we've included a massive glossary at the back of the book, starting on page 219, that probably covers far more ham terms than you will ever need to know.

No matter what your level of experience, ham radio always involves overcoming some significant technical barriers – really, that's what ham radio is all about. These barriers can range from the "simple" process (note those quotation marks …) of programming a local repeater's frequencies and tones into a handheld transceiver to building an HF station. I can't solve *all* of those challenges for you, but I'll give you some ways to think about the problems, and some resources.

The book is arranged by frequency bands, because each band brings its own equipment requirements, operating procedures, and modes of operation. We'll start with the most easily accessible bands, VHF/UHF, then work our way through MF/HF. While there's some interesting experimentation going on in our newest bands, the Super Low Frequency (SLF), 136 kHz (2,200 m) band and the Ultra-low Frequency (ULF), 472 kHz (630 m) band, those bands are a bit out of reach for most folks just beginning their ham careers. Likewise, the frequencies above the 70 cm band require specialized equipment and are not anywhere near as popular as VHF, UHF, MF, and HF; Very High Frequency, Ultra High Frequency, Medium Frequency and High Frequency.

This book is not a comprehensive guide to every piece of equipment and mode of operation. In fact, it's almost the opposite; it's easy to get lost in the ham radio equipment catalogs and other references, so I've kept things as simple as possible.

You should know, I probably have a more casual attitude toward most of this stuff than some hams out there. If you're a perfectionist about ham radio who must have everything technically perfect before you proceed, you're probably never going to work up the nerve to push the button and start making contacts. You need to get this in your bones:

Everything you will ever do in ham radio – *everything* – **is an experiment.**

You can't afford the kind of engineering and equipment that would make everything work perfectly the first time every time. Not even professional telecom companies and broadcasters get those kinds of results!

You also won't make it through your first hour or two on the air without pulling some hilariously boneheaded move, like forgetting your call sign at the most inappropriate possible moment, hopelessly scrambling the NATO Phonetic Alphabet, or one of countless other possibilities. A year from now you'll laugh about it, but why wait? Start laughing now!

Let us cast aside perfectionism, embrace the adventure, and boldly radiate!

"I Have no Idea How to Build a Station"

If you hold a Technician Class license, you have probably realized by now that the Technician license exam doesn't cover very much in the way of "nuts and bolts." If you're a General, you know the General doesn't cover much of that, either and, wouldn't you know it, neither does the Extra.

Ideally, you learned the fundamentals of radio as you progressed to your license, but it's still a long leap from Ohm's Law to a working radio station. After all, it took 27 years to get from Ohm to Heinrich Hertz's demonstration of radio waves, and another 13 years before Guglielmo Marconi announced he had achieved trans-Atlantic wireless communication.

It wouldn't be fair to blame the exam committees for this gap between theory and practice. The topics of the exams are dictated by the FCC, the Federal Communications Commission, and the exams have *always* been long on theory and short on practice. Speaking as someone who works full-time at teaching new hams the exam material, I'm not at all sure it would be practical to include much more how-to, because there are so many different "how-to's." The idea of the exams is to assure you have a context of knowledge that will allow you to experiment with and learn the practical stuff. Think of all that theory as the garden soil in which your practical knowledge will grow.

There was a time when new hams were usually taken under the wing of a more experienced ham. Those experienced hams are known as "Elmers" and "Elmas" and "Elmering or Elmaing" was a regular part of the hobby. It was an honor to be someone's Elmer.

For reasons I confess I do not know, that practice has diminished. That's why this book exists – consider Kerry and me your Elma and Elmer. Much like the Elmers and Elmas of days gone by, we have a limited number of stations to show you, and a limited number of ideas to share: however, our intention is to show you enough "attach part A to brace C" stuff that you'll be able to start building your own station(s) with some confidence that you're proceeding in a useful direction.

Every amateur radio station contains three fundamental parts; a transceiver, a feed line, and an antenna. In a handheld transceiver, those are all in one box and all installed by the manufacturer, but they're still present. For stations that aren't handheld, it will be your job to obtain those items, then hook them together properly. Auxiliary equipment for a station might include a power supply for the transceiver, an antenna tuner, and various other gear related to safety and/or convenience. We'll cover those as we go along.

We're going to take you through the builds of three distinct stations; a mobile VHF/UHF station, a fixed VHF/UHF station, and an HF station. We urge you to think of these stations as examples – prototypes, one might say. Along the way, we'll be explaining the principles behind how we're putting the station together, and those principles are what we really want you to learn from this journey. There are people in the world who are great at following recipes but might starve to death without a cookbook; then there are folks who deserve the title Chef; they know about esoteric stuff like thermal mass and Maillard reactions, and can cook great food without ever looking at a recipe – but those people *started* cooking by following recipes, too. My advice is start by following the "recipes" in this book with an eye toward eventually becoming an amateur radio "Chef".

"There Are So Many Rules! I'm Sure to Break One!"

You would think, with all the emphasis the rules and regulations get in the license exams, that we'd all be experts on the subject. A quick scan of just about any internet amateur radio group reveals this is very definitely not the case.

It's surprising to me how few hams have ever read FCC Part 97. It's not a long document – only about 34 8½ x 11-inch pages. You don't need to be a lawyer to read it; it is written in reasonably plain language. It's even free! You can download it from:

https://www.govinfo.gov/content/pkg/CFR-2009-title47-vol5/pdf/CFR-2009-title47-vol5-part97.pdf

Fast Track Ham Radio Facts contains a copy that is indexed for easy reference.

I think it is because we are, in general, a law-abiding bunch of folks that in the absence of a firm grasp of the rules and regulations, we feel less than 100% confident about what we should or should not do on the radio. Sometimes, *we even make up imaginary rules*! I followed a long and painful-to-behold thread unfold on one ham radio site about the timing of station identification. The amount of misinformation presented as fact was genuinely breathtaking. It was also hundreds of times longer than the full text of the rules regarding the topic:

(a) Each amateur station, except a space station or telecommand station, must transmit its assigned call sign on its transmitting channel **at the end of each communication, and at least every 10 minutes during a communication**, for the purpose of clearly making the source of the transmissions from the station known to those receiving the transmissions. No station may transmit unidentified communications or signals, or transmit as the station call sign, any call sign not authorized to the station.

That seems quite straightforward to me. By the way, I constantly hear people fail to identify their station every 10 minutes, and at least one member of our club has, so far as I know, *never* ID'd "at the end of each communication." Does anyone care? Nope.

The truth is, the FCC is not going to roll up in front of your house with a SWAT team to take you off to radio jail if you bobble a rule. They have neither the budget nor the interest to enforce the amateur radio rules, even in the case of a few blatant and repeated abuses of the airwaves. If you're making a good faith effort to operate your station "in accordance with good engineering and good amateur practice," and you're observing common courtesy, you're not going to have a problem. Almost certainly, the very worst that would happen is a warning letter from the FCC – and those are very, very rare. In the very unlikely event you should get one, just comply with the letter, notify them that you have done so and will continue to do so, and all will be well.

The FCC is also almost certainly not going to pay you an unexpected visit and inspect your station, despite what that question about this on the Technician exam says. I spent almost a quarter-century working full time in *commercial* broadcasting – where the FCC is a *lot* more concerned with station operations – and I never saw a single FCC inspector. (One did visit our station for about 30 minutes once while I was out to lunch. Nothing bad happened.)

It seems to me that the exam process has an unanticipated outcome: It leaves a lot of people feeling like the title of this section; "There are so Many Rules! I'm Sure to Break One!"

Let's go over the stuff you must do and the stuff you must not do. Table 1.1, on page 12, shows the rules that apply to what I'd call "normal", day-to-day ham radio operations, and there aren't very many of them.

Then there are topics that come up very rarely for most of us. When they come up, there's no need to guess; turn to your handy copy of Part 97 and look up the rules.

Table 1.2 is a list of events that should tell you, "I should take a look at Part 97 before I charge ahead blindly."

Finally, there's that whole business of RF (radio frequency) exposure evaluations. This is not complicated, and it is something you'll want to do, if only for your own peace of mind.

Once you know where to find help on this, doing your RF exposure evaluation is no more difficult than Googling something. First, consult the chart shown in Table 1.3 (page 14).

Let's say you're planning to operate on the 17-meter band. The chart says if your PEP is less than 125 watts, no formal evaluation is required.

But let's say you're planning to operate on the 2-meter band at 100 watts. Uh oh – on the 2-meter band you can only go up to 50 watts before an evaluation is required. Here's how you do that. You go to this web site:

http://hintlink.com/power_density.htm

The web page asks for the power *at the antenna*. Since that's guaranteed to be less than your transmitter's output power, just use that, at least for your first pass. Then you enter the antenna gain in dBi (you can look it up by finding your antenna on the internet) the distance to the "area of interest", which is where the closest people are, and the frequency of operation. Click on "Calculate RF Power Density" and you'll have your evaluation instantly.

Figure 1.1 shows the screen where you enter your data. I've plugged in my transmitter's full power, 100 watts, and the distance to the closest place where we use the radio. That would represent the "controlled area", where our exposure is allowed to be higher than in, say, the neighbors' house, which is an "uncontrolled area." That happens to be a good deal farther away from the antenna than our operating position.

Once you click the "Calculate RF Power Density" button, you'll be taken to a

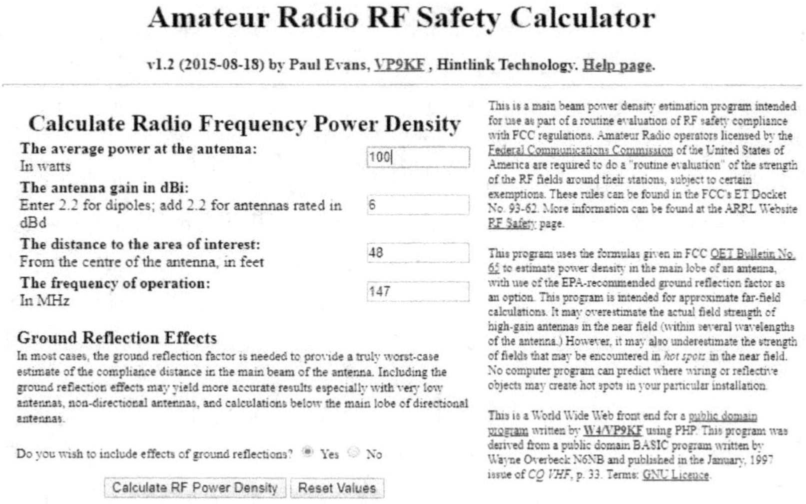

Amateur Radio RF Safety Calculator

v1.2 (2015-08-18) by Paul Evans, VP9KF, Hintlink Technology. Help page.

Calculate Radio Frequency Power Density

The average power at the antenna: 100
In watts

The antenna gain in dBi: 6
Enter 2.2 for dipoles; add 2.2 for antennas rated in dBd

The distance to the area of interest: 48
From the centre of the antenna, in feet

The frequency of operation: 147
In MHz

Ground Reflection Effects

In most cases, the ground reflection factor is needed to provide a truly worst-case estimate of the compliance distance in the main beam of the antenna. Including the ground reflection effects may yield more accurate results especially with very low antennas, non-directional antennas, and calculations below the main lobe of directional antennas.

Do you wish to include effects of ground reflections? ● Yes ○ No

[Calculate RF Power Density] [Reset Values]

This is a main beam power density estimation program intended for use as part of a routine evaluation of RF safety compliance with FCC regulations. Amateur Radio operators licensed by the Federal Communications Commission of the United States of America are required to do a "routine evaluation" of the strength of the RF fields around their stations, subject to certain exemptions. These rules can be found in the FCC's ET Docket No. 93-62. More information can be found at the ARRL Website RF Safety page.

This program uses the formulas given in FCC OET Bulletin No. 65 to estimate power density in the main lobe of an antenna, with use of the EPA-recommended ground reflection factor as an option. This program is intended for approximate far-field calculations. It may overestimate the actual field strength of high-gain antennas in the near field (within several wavelengths of the antenna.) However, it may also underestimate the strength of fields that may be encountered in *hot spots* in the near field. No computer program can predict where wiring or reflective objects may create hot spots in your particular installation.

This is a World Wide Web front end for a public domain program written by W4/VP9KF using PHP. This program was derived from a public domain BASIC program written by Wayne Overbeck N6NB and published in the January, 1997 issue of *CQ VHF*, p. 33. Terms: GNU License.

Figure 1.1: RF Exposure Calculator Data Entry Page

page like Figure 1.2.

As you can see, we're well within compliance at 100 watts.

Be clear; there *is* more to Part 97 than we have covered here, and the rules make very clear that you are responsible for following all the rules, whether they were on one of the exams or not. However, there's also a lot more to the motor vehicle code than anyone who isn't in law enforcement or carrying a commercial driver's license knows, too. Chances are good, though, that you needn't be concerned about the many rules covering oversized truck loads in your state, nor are you likely to run afoul of the obscure rules about operating on the 23-cm band in certain rather empty regions of Colorado and Wyoming.

Amateur Radio RF Safety Calculator

Calculation Results

Average Power at the Antenna	100 watts
Antenna Gain in dBi	6 dBi
Distance to the Area of Interest	48 feet 14.6304 metres
Frequency of Operation	147 MHz
Are Ground Reflections Calculated?	Yes
Estimated RF Power Density	0.0379 mW/cm^2

	Controlled Environment	Uncontrolled Environment
Maximum Permissible Exposure (MPE)	1.005 mW/cm^2	0.205 mW/cm^2
Distance to Compliance From Centre of Antenna	9.3933 feet 2.8631 metres	20.9423 feet 6.3832 metres
Does the Area of Interest Appear to be in Compliance?	yes	yes

Interpretation of Results

1. The power value entered into these calculations should be the average power seen at the antenna and not Peak Envelope Power (PEP). You should also consider feedline loss in calculating your average power at the antenna.

2. If you wish to estimate the power density at a point below the main lobe of a directional antenna, and if the antenna's vertical pattern is known, recalculate using the antenna's gain in the relevant direction.

3. Please also consult FCC OET Bulletin 65 Supplement B, the Amateur Radio supplement to FCC OET Bulletin 65. It contains a thorough discussion of the RF Safety regulations as they apply to amateur stations and contains numerous charts, tables, worksheets and other data to help determine station compliance.

Perform another computation

Figure 1.2: RF Exposure Calculator Results Page

Topic	You Must	You Must Not
Station Identification	ID once every 10 minutes and at the end of the communication.	Send unidentified communications or use a false call sign.
Transmit Frequency	Only transmit within the ham bands. If you are near the top or bottom of a band, remember to take into account the bandwidth of your signal.	Intentionally interfere with any legal transmission, including those of other hams.
Transmitter Power	For most bands, licenses and modes, limit your power to 1500 watts output from the transmitter. Novices and Technicians must not exceed 200 watts PEP on HF bands. Geographical power restrictions apply to the 630-meter, 70-centimeter, 33-centimeter and 23-centimeter bands.	Exceed the legal power for the band and/or mode on which you are operating, nor exceed the maximum power necessary to carry out the intended communication.
60 Meters	Limit your power to 100 watts Effective Radiated Power relative to a dipole.	Transmit on 60 meters outside the five channels. See separate table.
30 Meters	Limit your power to 200 watts PEP.	Transmit anything but CW, RTTY, and data on 30 meters.
Transmission Content	Limit international transmissions to "communications incidental to the purposes of the amateur service and to remarks of a personal character." (Don't use ham radio to foment a revolution in a foreign country.)	Transmit music, obscene or indecent language, or messages from which you will profit, directly or indirectly, except in special circumstances; you may advertise ham radio gear for sale so long as you do so on an occasional basis, and you are not a professional dealer.

Table 1.1: Part 97 Musts and Must Nots

Topics to Look up as Needed		
Spread Spectrum	Traveling out of the country	Controlling model craft with amateur radio
Putting up a repeater	Operating on 70-cm near Canada.	Putting up an automatically controlled digital repeater
Third-party International Communications	Renewing your License	Operating in bands outside the range of 160-meters through 70-cm.
Interference Complaints	Re-transmission of signals from other stations, such as weather stations.	Change of Address
Putting up a Tower	Putting up an auxiliary station	Operating near a radio-telescope facility.
And any time you get that, "Is this right?" feeling!		

Table 1.2: "I should look this up ..."

RF Power and Evaluation Requirements	
80, 75 & 40 meters	500 watts
30 meters	425 watts
20 meters	225 watts
17 meters	125 watts
15 meters	100 watts
10 meters	50 watts
6, 2, and 1.25 meters	50 watts
70 cm	70 watts
33 cm	150 watts
23 cm	200 watts
13 cm and higher	250 watts
Repeater stations on all bands	Non-building mounted antennas: Height above ground level to lowest point of antenna less than 10 meters (32.8 feet) *and* power greater than 500 watts *ERP*. Building-mounted antennas: Power greater than 500 watts *ERP*.

Table 1.3: RF Exposure Evaluation Requirements

"I'm a Little …uh …Mic Shy…"

We'll hazard a guess that this is the real number-one stumbling block for the new ham. Maybe you don't feel this way at all. If so, we envy you! If you do experience at least a little nervousness about getting on the air, the first thing we want you to know is that you are far from alone.

Contemplating speaking on a microphone into a transmitter can feel a lot like contemplating public speaking. Everyone's going to hear you, right? Studies repeatedly show that when people rank their fears, fear of public speaking ranks as scarier than death! Yes, what many people say is that they would rather die than speak in public. I've always thought that if their real choice was to speak in public or die immediately, they'd be on the podium in a heartbeat, but I understand very clearly what they're saying.

What are we afraid of? Well, we're afraid we'll make utter fools of ourselves, of course! We won't know what to say, or we'll say the wrong thing and be laughed at or scolded. Our voice will sound funny. We'll freeze! We'll panic and run down the street clucking like a chicken! It's all ridiculous, I know, but even though we're normally functioning adults, if you put a live microphone in front of us, all of a sudden, we're six years old again. I don't know about you, but I can't think of many six-year-olds I want making my life decisions, but if I don't overcome this internal conversation, that's exactly what happens.

I don't have a miracle cure for this feeling, either, much as I wish I did. I've been appearing on stage, and on radio and television since I was 5 years old, and every time I make a speech, do a presentation, start a class, make a toast at dinner, or even think about going to a party, I go through the same nervousness. It never seems to get better. It's just part of the deal, I guess, because I've never known anyone for whom it got better.

Goodness knows I tried lots and lots of the "solutions" in the self-help section of the bookstore, with zero (or worse) results. What does work, though, is simply making "doing this thing" more important than the nervousness. With that mindset and practice, the nervousness becomes less meaningful. It's present, but it doesn't mean "quit," it just means I'm nervous.

Most important, the nervousness is always temporary. By that, I mean it evaporates once I'm making the speech, leading the class, or doing whatever it was that looked so daunting before I started. You've probably had the same experience.

The good news about feeling a little nervous about talking on the radio is that it shows you care about the hobby and about being excellent at the techniques of the hobby. That's great! You might consider backing off on the perfectionism as you get started. Nobody – seriously, *nobody* – starts any new endeavor as a master. Mastery takes time, and it takes at least a little capacity to embrace failure.

Wait – did I say *embrace* failure? Isn't failure something to be avoided at all costs? That's what our ego says, but just a little thought will reveal that the overwhelming majority of successes are preceded by failures.

Let's see if we can get a little perspective on this issue.

There isn't a giant crowd out there. A really big net might consist of 50 people. More likely a bare handful of people will hear you – if that. We're in an active area for ham radio, but there have been many times I've gotten in the car, switched on the radio, said, "AF7KB monitoring …" and heard nothing but radio silence for the next twenty minutes; and that's on VHF/UHF. Wait till you spend 20 minutes calling CQ CQ CQ this is AF7KB CQ CQ CQ on the HF bands and get no replies! Then you'll really understand that at any given moment, the "audience" for your station is pretty small.

Even if you stumble into that net with 50 participants, it's still not a crowd. They're just 50 individuals – so you don't need to speak as if you were speaking to a crowd. Just speak to one person.

It helps if you imagine that single person to whom you're talking is very friendly. I know there are a few sourpusses out there who think they are the World's Greatest Authorities on How All Things Should be Done and are quick to criticize anyone who doesn't live up to their (completely fabricated) standards. These windbags often are quick to announce that, by cracky, when they got *their* license, they had to walk to an FCC office 100 miles away through the snow – uphill! Both ways! They'll tell you the hobby has done nothing but gone downhill "since they dropped the Code Requirement." (You can tell Code Requirement is capitalized when they say it.) You, of course, are Part of the Problem. As Part of the Problem, you're in excellent company, by the way; based on a few hamfest encounters, I'm a Big Part of the Problem because of the titles of our books. *Fast Track* indeed! Harumph!)

There is absolutely nothing you can do to avoid these curmudgeons' judgments. Even if you had a time machine to transport yourself back to an earlier time so you could get your license "back in the day", they'd gripe about your operation of your time machine.

So what?

In my opinion, they've completely lost sight of the purpose of the hobby. Ham radio isn't an exercise in doing a bunch of arbitrary procedures correctly, nor about having the proper (in their imaginations) equipment. It's about playing with technology and connecting with people. All the procedures and equipment exist solely to facilitate those.

Happily, these grouches constitute a very small percentage of the hams out there – they just happen to be a loud percentage.

In my experience, the vast majority of hams are extraordinarily gracious and polite. They're also happy to welcome newcomers to the hobby and don't care one bit if you say "seventy-three", "seventy-threes" or "seven-three." Pay attention to those folks and ignore the negative ones.

Despite the impression left by the exams, ham radio is mostly conducted in ordinary English, especially on VHF/UHF. There are a few basic, common sense procedures to know that are designed to keep communications clear, and you'll catch on quickly enough to those just by listening, but no one cares if you say "static" instead of "QRM."

But, what does one talk about? That's one of the beauties of ham radio! I am almost completely inept at the art of making small talk. Plop me down in a room full of strangers with no common interests and I do not thrive. On the ham radio, though, there's always some ham radio related topic. What sort of radio are you using? Where are you? How's my signal? Oh, you're using that new Super-Duper Multi-Band 101X antenna? How's that working for you? How high do you have it mounted? It goes on and on.

The rewards of being an active ham are many; you'll make amazing, generous friends. You'll have adventures. You'll make great memories, and you'll have some fun. The only price for all those rewards is working up enough nerve to push that button, say your call sign, and see what happens – which happens to be what everybody out there is waiting for you to do.

One last thought. There is no doubt that it is possible to overcome "mic shyness", "first-time jitters", "stage fright", or whatever else you care to call it. Shyness is not a medical condition; it's just a state of mind. It can be overcome by a different state of mind.

We have a friend who was petrified – almost literally – not just by the prospect of speaking on the radio, but even by the prospect of getting her license. Let's call her Violet. By nature, Violet is a *very* introspective, quiet person. Her husband is an enthusiastic ham and an equally enthusiastic promoter of the hobby. He really wanted to share the hobby with his wife. The day came that she got up her nerve enough to go and get her Technician license – as I recall, she got 100% on the exam.

Nothing happened on the radio. A few months later, she passed her General exam. Still, she was very clear; she was *not* talking on the radio, thank you very much!

It was near this time that the co-author of this book, Kerry, KC7YL, started up our club's Thursday night YL net. With a lot of encouragement from her husband and from Kerry, lo and behold, Violet appeared one night on the YL net. Her appearance was brief; basically, [call sign] [name] [location] and, "I-don't-really-have-anything-to-share-this-is-[call sign]-back-to-net," but she talked on the radio! Then she became a regular on the net, and each week she'd open up a little more and sound a little more comfortable.

Now, she's studying for her Extra and has discovered a passion for SOTA – Summits on the Air – operations. She and her husband are out many weekends, scaling the many summits we have in Washington State and operating on VHF and HF from atop some of our beautiful mountains.

Believe me, if Violet can do this, you can do this; and when you do, your story will be someone else's inspiration to be brave and achieve something they want to achieve.

Part II

Getting Started on the VHF/UHF Bands

Chapter 2

Working Repeaters

I imagine there are a few reasons why the starting point for most of us is working repeaters on the VHF/UHF 2m/70cm bands – 144 to 148 MHz and 420 to 450 MHz.

First, VHF/UHF is where most of the Technician privileges are! Yes, Technicians have some HF (below 50 MHz) privileges, but as things stand now, those HF privileges are almost all CW only. The sole exception is the 10-meter band and 10-meter propagation has often been rather bleak here at the bottom of the sunspot cycle.

VHF/UHF is, overall, much easier to set up than HF. The entire act of creating a 2m/70cm "dual-band" radio station can consist of as little as switching on a handheld transceiver and programming the frequencies and tones for a repeater. Even fixed and mobile stations are simpler to create, in large part because VHF/UHF antennas are significantly smaller than antennas for the HF bands.

Propagation on the VHF/UHF bands is a much simpler proposition than HF propagation, too. There are some more esoteric forms of VHF/UHF propagation and more advanced ways of working those VHF/UHF bands, but no one talking on a local repeater is worried about the day's sunspot number; if I can hit our club repeater from my driveway tonight, it's 99% certain I'll be able to do the same tomorrow.

Dual-band radios and antennas are typically much less expensive than HF rigs. Amazon sellers will happily sell you a dual-band handheld unit for under $30. Despite what you may read on the internet, they're generally not horrible radios – I've worked a repeater that was about 40 miles away with one. (I cheated – I was on a hill, and that repeater is on a mountain.) You can get a nice handheld dual-band transceiver for under $100. Mobile/fixed transceivers from the major manufacturers are in the $300 to $400 range. The dual-band rig in our truck is *very* snazzy, with APRS (Automatic Packet Reporting System or Automatic Position Reporting System) and more features than I'll probably ever master, and even that rig would cost you less than $700, complete with antenna, feedline, and mount. That's a very low price for even a basic HF station, unless you're a hard-core bargain finder.

Of course, the drawback – if it is one – of VHF/UHF is that those bands are mostly local communications frequencies. However, that has been changing, with the advent of internet-linked technologies such as EchoLink, IRLP (Internet Radio Linking Project), and the various digital voice modes such as Fusion (also known as C4FM), D-Star, and DMR. (I realize there are some hams who think any communication involving anything besides over-the-air transmission by electromagnetic waves is "not real radio." I see their point and they're welcome to it, but I think it's costing them some fun.)

Finding Your Local Repeaters

To work a repeater, you need to know the repeater's *transmit frequency* – that's the frequency on which the *repeater* transmits, not the one on which you'll transmit. If a ham tells you "most of us are on the 147.180 repeater," that means to hear that repeater, you listen on 147.180 MHz.

You also need to know the repeater's *offset*. That will tell you the frequency on which you'll transmit. In the case of the 147.180 repeater in this example, the standard offset is +0.6 MHz, so you'll transmit on 147.780. (There are rare exceptions to those standard offsets. You are unlikely to encounter them.)

Finally, you'll need to know the repeater's *tone*. The "tone" is a low audio frequency sent by your radio along with your voice. If you try to reach a typical repeater without transmitting that tone, the repeater will just ignore your transmissions; the transmitter side of the repeater won't switch on. In the vast majority of cases, the tone will be a CTCSS (Continuous Tone Coded Squelch System) tone, but a few repeaters use what's called a DCS (Digital Code Squelch) tone. CTCSS tones have a decimal point in them, such as 127.3 or 77.0. DCS tones do not contain a decimal point and are usually designated by a "D" at the start, such as D026. In either case, the tone is what tells the repeater, "This transmission is intended for this repeater." Without the tone squelch system, the repeater listening to 147.780 would just repeat everything it heard transmitted on 147.780, whether it was intended for that repeater or not. Using the tone system lets us put more repeaters into a limited amount of geographical space and electromagnetic spectrum.

How does one come to know all those frequencies and tones?

One time-honored way is to order a paperback Repeater Directory from the ARRL.

Another is to visit the very well-organized website called:

http://repeaterbook.com

The information at repeaterbook.com is also available in app form for your smart phone or tablet, but there are lots of features on the desktop version that are not, as yet, available on the app.

You can search repeaterbook.com in a number of ways, including by US state and city, by wavelength, and, in case you're planning a road trip, by major US freeway.

Figure 2.1 (page 23) shows the repeaterbook.com listing for our local club repeater.

Everett, WA

WA7LAW
Repeater ID: 53-127

REPORT UPDATE

Downlink:	147.18000
Uplink:	147.7800
Offset:	+0.6 MHz
Uplink Tone:	103.5
Downlink Tone:	103.5
County:	Snohomish
Call:	WA7LAW
Use:	OPEN
Op Status:	On-Air
Sponsor:	Snohomish County Ham Radio Club
Features:	300 watts, closed autopatch, e-power.
Mixed-Mode:	Yes; analog capable.
Nets:	WA7LAW: Sun at 19:00.
Web links:	http://www.wa7law.com/wa7law/repeater.html

Last update: 2018-04-22

Figure 2.1: RepeaterBook Listing for WA7LAW

As you can see, the listing tells me to listen on 147.180, and transmit on 147.780. The "uplink tone" is 103.5 Hz. That's the "tone" that will switch on the repeater when we transmit. Our repeater also sends a "downlink tone", also at 103.5 Hz. That downlink tone operates like the uplink tone, but in reverse; it switches your radio's receive function on if the radio is set up to use downlink tone squelch. If I wanted to, I could set my radio to only respond to signals on 147.180 that included that downlink tone. This downlink tone function is something that has created a lot of frustration and confusion for many beginners, since it is never mentioned in any of the ham exams. The result of setting a value for the downlink tone is usually silence, because most repeaters do not send a downlink tone; if you have downlink tone squelch set, it will seem as if the repeater is not transmitting because your radio is remaining silent while it awaits (forever) a signal that includes that downlink tone. My recommendation is to *not* enter any value for that downlink tone. That way your radio will hear everything.

To make downlink-tone-related matters just a little more confusing, "downlink tone" goes by various names, depending on the radio and/or programming software you are using. The popular free programming software known as CHIRP, for instance, calls it "ToneSql" for "tone squelch." The Baofeng UV-5R instructions call it "R-CTS" for "Reception Continuous Tone Coded Squelch." Once you understand the concept, though, you'll figure out the proper course of action.

Chapter 3

Dual-band Handheld Transceivers

The starting point of lots of hams' radio collection is a dual-band handheld transceiver, an HT. I think most hams have at least one stashed somewhere, and that includes a lot of hard-core HF folks.

Figure 3.1: Yaesu FT3DR Handheld Transceiver

Handhelds are great for the purpose for which they were designed. As stand-alone devices, though, HT's are limited.

They are low-power radios, typically 5 watts, but their real limitation is their antenna. The laws of physics say it's just not possible to create a high-efficiency, high-gain, low-loss, wide-bandwidth antenna that's easily portable. Beyond that,

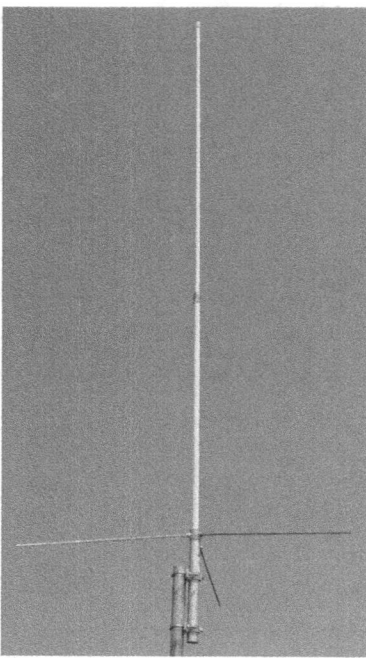

Figure 3.2: Tram Amateur Dual-Band Base Antenna

antennas work best outdoors, which restricts the HT's usefulness indoors. There's not a single handheld in our home that can hit our club's repeater – or any other repeater, for that matter – from inside the house. We can hear our local repeater, but can't get enough signal out to be heard. Friends who live much closer to the repeater can make their handhelds work, but here at the house, our handhelds just do duty as scanning receivers.

It is possible, though, to remove the "rubber duckie" antenna and, with the appropriate adapter, attach an exterior antenna. This can be as simple as a mobile magnetic mount antenna or a full-scale, permanently mounted dual-band antenna such as the Tram unit in Figure 3.2.

That Tram antenna stands about eight feet tall and mounts to masts 1.81 in. to 2.44 in. in diameter. It has a power handling capability of up to 200 watts, and 6 dBd of gain on VHF and 8 dBd on UHF. Curiously enough, it is available through Home Depot – not a place one usually expects to find specialized ham gear! The design of the antenna is "colinear." That's not a type of antenna covered on the ham exams; inside that weatherproof shell, it's essentially four half-wave 2-meter dipoles stacked vertically, with tuned radials at the base.

Figure 3.3: Baofeng UV-5R

For mobile use, you can stick that mag-mount antenna on the roof, bringing the feed line through a door – the foam weather stripping should prevent it from being crushed – and have a perfectly serviceable mobile radio. Mobile installations don't get any simpler than that.

If you're using a magnetic mount antenna at home, you can stick it to any handy large sheet of steel, the bigger the better. Even a cookie sheet will do in a pinch! The sheet metal is an important component of the antenna; it serves as a ground plane, which almost every mag-mount requires. (A ground plane acts something like an electrical mirror, effectively doubling the length of the antenna.) In normal use, of course, the ground plane is the roof of your car, so you need a substitute "car roof" to make it work.

Figure 3.3 shows one of those $30 dual-band handheld transceivers I mentioned.

That's a Baofeng UV-5R – sometimes marketed as a "Pofung UV-5R" or "BTech UV-5R." It's enormously popular, especially among new hams.

Those are typically packaged with a few accessories that range in value from "really must have" to – frankly – "really should just throw away." The must-haves are the charger stand and a programming cable that allows you to connect the radio to your computer for the purpose of programming in all the memory channels. As you'll see, this is very desirable compared to programming it manually. The little external microphone/speaker is on the "really should just throw away" list, although there are YouTube videos showing modifications you can make to it that claim to improve it. (Nothing could make it worse...)

I doubt that little radio is very durable. The durability of handhelds gets tested if you're planning to walk around with the thing clipped on your belt since it gets bounced and jostled with every step. I've never dropped mine on a hard surface, and don't plan to, but I'm doubtful about its ability to survive. It is certainly not waterproof, nor even highly water resistant. The supplied antenna is not the greatest; at least, that's the general belief in the ham community. I've never seen any scientific testing on the topic. There are upgraded antennas available that are quite inexpensive.

The FCC is not fond of certain Baofeng UV-5R's because they are capable of transmitting on all sorts of frequencies – including the ham bands – that are not covered by their "type acceptance." You may have read that they are illegal. Certain models have been banned from being imported. They are *not* illegal for hams to use *on ham frequencies* and at least two FCC spokespeople have announced exactly that, for the public record.

Let me be clear. I am not endorsing this radio, even though I own a couple. They are what they are – an inexpensive, "not-too-bad" handheld radio. I consider them to be my disposable radios. If we're going out for a walk and the weather looks dicey, but, "Hey, let's keep an ear on the repeater," the Baofengs get the nod. When I'm serious about communicating, they won't be the ones I choose. Baofengs are so widely owned, though, it would be hard to leave them out of this book.

Manually Programming a Baofeng UV-5R

Before I launch into this, I really want to tell you not to panic! There are at least two far, far simpler ways of doing this with any modern dual-band radio, and I'll show you those ways. Radios made by Yaesu, Kenwood, Icom, and Alinco are generally much more intuitive to program than the UV-5R. When I got a new Kenwood mobile radio a few months ago, I learned to program it in less than a minute.

Let's face it, though, even for the major brands, there's a limited amount of real estate on any handheld for buttons and display. By necessity, the buttons must be multi-purpose and displays can only display so much.

Somewhere out in the future, you might want to take on the task of programming a station into a Baofeng manually.

Unfortunately, at least to my mind, there is absolutely nothing intuitive about programming this Baofeng radio. It is almost as if the engineers set out to make it as utterly confounding as possible. The exterior of the radio offers you no clear clues about how to proceed, nor does the very limited screen. The "helper voice" built into it is heavily accented and sounds as if the main design objective behind it was to use as little memory as possible.

The included instructions – which I notice seem to have been vastly improved from how I remember the originals – still offer merely the faintest clues; they only list the functions of each control. There's no step-by-step "how-to-achieve-the-one-thing-every-user-will-want-to-achieve."

In short, this thing qualifies as user-hostile.

Despite all this, it is possible to get the job done in a mere 14 steps, which you can see in Table 3.1.

	Table 3.1: Programming a Baofeng UV-5R
Step 1	Press **[VFO/MR]** and enter *Frequency Mode*. **[VFO/MR]** switches between "Variable Frequency Oscillator mode" and "Memory Recall mode." In VFO mode, the up/down arrow buttons change the frequency up or down. In Memory Recall mode, the up/down arrows scroll through the programmed memory channels. The UV-5R has 128 memory slots. Most other dual-band radios have a similar memory feature – some have many more memory slots.
Step 2	The UV-5R display shows two frequencies, one atop the other. They are referred to as the *A side* and the *B side*. The A side is the "side" on top. Press **[A/B]** and choose the *A Side* (upper display). The A side must be used to program the repeater channels into the radio. You can enter programming data on the B Side (lower display) but it will not be saved. You won't get a warning message about this, either.
	Continued on next page...

	Table 3.1: Programming a Baofeng UV-5R	
Step 3	Press **[BAND]** for the frequency band. Toggle **[BAND]** to choose 136 MHz (VHF) or 470 MHz (UHF). If the incorrect band is chosen for the frequency entered in Step 6, the radio will cancel the operation.	
Step 4	*(Optional.)* Clear any CTCSS/DCS codes previously assigned to the channel. If no previous codes exist or when setting up the channel for the first time and no codes are needed, set the menu items listed below to OFF. *RX DCS* - **[MENU]** 10 **[MENU]** (enter **0** (OFF)) **[MENU] [EXIT]** *RX CTCSS* - **[MENU]** 11 **[MENU]** (enter **0** (OFF)) **[MENU] [EXIT]** *TX DCS* - **[MENU]** 12 **[MENU]** (enter **0** (OFF)) **[MENU] [EXIT]** *TX CTCSS* - **[MENU]** 13 **[MENU]** (enter **0** (OFF)) **[MENU] [EXIT]**	
Step 5	Disable TDR (*DualWatch/Dual Standby*). Press **[MENU]** 7 **[MENU]** (press up/down arrow keys) OFF **[MENU] [EXIT]** It is highly advised to turn TDR off when programming directly from the radio.	
Step 6	*(Optional.)* Delete any existing data on the channel to program. Skip this step when setting up the channel for the first time. Press **[MENU] 28**. Press **[UP/DOWN ARROW KEYS]** to choose channel number **[MENU] [EXIT]** It is highly advised to turn TDR off when programming directly from the radio.	
Step 7	Enter the repeater output (your receiving) frequency. Use the keypad to enter the frequency into the radio.	
	Continued on next page...	

	Table 3.1: Programming a Baofeng UV-5R
Step 8	Input the repeater frequency offset. Press **[MENU] 26 [MENU]** (enter the offset for 2 meter or 70 cm repeater) **[MENU] [EXIT]**
Step 9	Enter the Transmit Frequency Shift. Press **[MENU] 25 [MENU]** (enter **1** for positive shift or **2** for negative shift) **[MENU][EXIT]**
Step 10	*optional* - Enter the transmit CTCSS/DCS code. *CTCSS* - **[MENU] 13 [MENU]**. Enter/choose code XXXX. **[MENU] [EXIT]** *DCS* - **[MENU] 12 [MENU]** Choose code XXXXX. **[MENU] [EXIT]**
Step 11	Assign the receive frequency entered in Step 7 to the channel. **[MENU] 27 [MENU]** (enter channel number XXX) **[MENU] [EXIT]**
Step 12	Press the **[*Scan]** button to activate *Reverse Mode* and display the transmit frequency. (In normal use, the Reverse Mode is used to listen to the input frequency of the repeater. Sometimes you can hear a station better that way.)
Step 13	Assign the transmit frequency to the channel. Press **[MENU] 27 [MENU]** (enter the same memory channel entered in step 12) **[MENU] [EXIT]**
Step 14	Press the **[*Scan]** button to exit *Reverse Mode*.

Table 3.1: How to Program a Baofeng UV-5R

Nothing to it!

(Thanks to the folks at Buy Two Way Radios for their step-by-step. At the moment they feature a complete "ham radio starter kit" of a UV-5R with top quality accessories.

https://www.buytwowayradios.com/btwr-essentials-gsk-ham-uv5r.html)

Programming a UV-5R With a Computer

Feeling a little discouraged about programming a UV-5R manually? If so, I can't blame you. The UV-5R isn't a bad little radio, but that user interface is pretty dreadful.

I promised two alternatives, and here they are.

These both require a programming cable for your radio. It will have a USB plug on one end and, in the case of the Baofeng UV-5R, a connector that fits into the "remote mic/speaker" jacks on the radio.

The cable, shown in Figure 3.4, allows your computer to communicate with your radio. There is a chip behind the USB plug that serves as the translator between your computer and the radio.

Figure 3.4: Baofeng Programming Cable

You'll probably get a mini-CD of programming software with your radio. If not, you could download it from Baofeng, but don't bother. In fact, it's best not to remove the CD from its envelope before humanely disposing of it, so there's no danger of you ever installing it. The same people designed the software as designed the interface on the radio! My last encounter with it led to nothing but sadness.

Instead, download a free program known as CHIRP. It's available from

http://chirp.danplanet.com

CHIRP lets you program all the memories on a layout something like a spreadsheet. You download your radio's current memory configuration, alter it to your needs on the computer, then upload that configuration to the radio. *Much* simpler than all the [MENU] 12 [EXIT] jazz. Figure 3.5 (page 33) a screenshot of CHIRP in action that shows the first ten channels I have programmed into my radio today:

The column marked "Loc" is the memory slot number. The "Frequency" col-

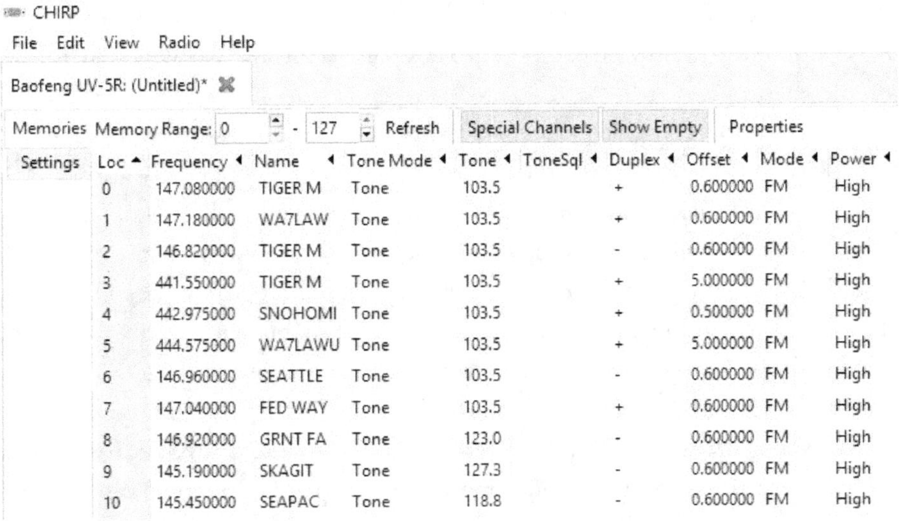

Loc	Frequency	Name	Tone Mode	Tone	ToneSql	Duplex	Offset	Mode	Power
0	147.080000	TIGER M	Tone	103.5		+	0.600000	FM	High
1	147.180000	WA7LAW	Tone	103.5		+	0.600000	FM	High
2	146.820000	TIGER M	Tone	103.5		-	0.600000	FM	High
3	441.550000	TIGER M	Tone	103.5		+	5.000000	FM	High
4	442.975000	SNOHOMI	Tone	103.5		+	0.500000	FM	High
5	444.575000	WA7LAWU	Tone	103.5		+	5.000000	FM	High
6	146.960000	SEATTLE	Tone	103.5		-	0.600000	FM	High
7	147.040000	FED WAY	Tone	103.5		+	0.600000	FM	High
8	146.920000	GRNT FA	Tone	123.0		-	0.600000	FM	High
9	145.190000	SKAGIT	Tone	127.3		-	0.600000	FM	High
10	145.450000	SEAPAC	Tone	118.8		-	0.600000	FM	High

Figure 3.5: CHIRP Programming Screen

umn is the transmit frequency of the repeater. Baofengs allow you to give each repeater a name up to seven characters long.

"Tone Mode" is selected from a pull-down menu.

- **(None)**: No tone or code is transmitted, receive squelch is open or carrier-triggered – in other words, squelch opens when it hears a signal above a certain strength, regardless of whether or not that signal is carrying a tone.

- **Tone**: A single CTCSS tone is transmitted, receive squelch is open or carrier-triggered. The tone used is that which is set in the "Tone" column. This is far and away the mode you will use most.

- **TSQL**: A single CTCSS tone is transmitted, receive squelch is tone-coded to the *same* tone. The tone used is that which is set in the "ToneSql" column.

- **DTCS**: A single DTCS/DCS code is transmitted, receive squelch is digitally tone-coded to the *same* code. The code used is that which is set in the DTCS Code column.

- **Cross**: A complex arrangement of squelch technologies is in use. If you get advanced enough to use this, you'll be familiar enough with CHIRP to set this up properly.

You'll notice the ToneSql column is blank on all of my memories. If I set a tone in that column, my radio won't respond to any signals on that frequency that don't contain that tone.

If you look closely down the "Duplex" column, you can see the "+'s" and "-'s" that indicate the direction of the offset of the repeater. The software sets this to the most common values automatically when you enter the frequency in the "Frequency" column, and it also sets the offset frequency, but you can alter both if needed. If you set up the national calling frequencies, 146.520 MHz and 446.000 MHz in two of your memory banks, which I highly recommend you do, you'll need to manually make those memory slots "simplex" channels by selecting "None" from the pull-down in the "Duplex" column.

You want "FM" for the "Mode" column. The other setting is "NFM" for Narrow Band FM. Narrow Band FM is much more often used on the Family Radio Service and General Mobile Radio Service bands than in amateur radio.

I typically leave the power setting for all the channels at "High" – that's easy enough to change manually if I need to. (How much excess power could I possibly have at 5 watts?)

Once you have all the settings where you want them, you "Upload to Radio" and you're all set.

It all sounds very simple, and it often is. To me, it is certainly better than manually programming the radio, since I find that process so annoying I'd almost rather have no radio at all. The process can, however, get – for want of a better term – a little wonky.

First, there's the matter of the programming cable. They are made by several Chinese manufacturers, including some who use chips in the cables that sometimes don't act as you'd expect. (In fact, sometimes they're simply the wrong chips, period.) The fakes are not easy to spot, but you'll see symptoms. For instance, on my system, using my cable, updates to Windows 10 often put the interface between the radio and CHIRP into complete failure. To make it work again, I must run a little hacker program that replaces some newly updated driver with an old version of the driver. Apparently, this demonstrates that I am the not-so-proud owner of a fake Baofeng programming cable! If you get a bum cable, you can probably ship it back to China for a replacement, at a cost that exceeds the price of the cable, or just eat the loss.

There is a far better experience available for a modest increase in cost.

RT Systems is a Colorado based company that has been making radio programming software and interfaces for many years. In fact, they were the people who invented amateur radio programming software back in 1995. They offer programming software for some 40 brands of radios, covering about 170 models. While there is only one version of CHIRP that serves a wide array of different radios, each RT Systems software package is written specifically for that model of radio and comes to you with all the right default values already set to enable the transfer of information from radio to computer and back again.

The software itself is full of useful features. For instance, you can swap files with other hams who are using the software, regardless of their model of radio. Planning a trip to Abilene, TX? Get in touch with an Abilene ham and get their file, and you're all set up with the Abilene repeaters. Don't happen to have a ham friend in Abilene? No problem. The software can automatically connect to multiple sources of repeater information. You can quickly build a file of Abilene repeaters, or even one that will cover all the repeaters along your route – even if that route is not along major freeways.

Their RT interface cables are also custom designed and manufactured to work with their software and the radio in question, so everything works together right out of the box.

The RT Systems package also offers direct control of your radio. In other words, you could set up your radio on your kitchen table, hook it up to an outside antenna, connect it to your laptop running the RT software, and barely ever touch the radio again – all the functions except Push-to-Talk and on/off/volume would be controlled from your computer screen. This is a great feature with any radio but given the patience-testing nature of Baofeng manual programming, this is a real game changer.

The RT Systems package of software plus programming cable lists for $49.

https://www.rtsystemsinc.com/default.asp

You can see a video demonstration of many features of the system at

https://www.youtube.com/watch?v=CorLMz0JHac&feature=youtu.be

Before we leave the topic of handheld transceivers, be aware that it is possible to get a lot more capability for not a whole lot more money. While it is possible to spend hundreds on a unit such as Yaesu's FT-2DR (Figure 3.6, page 36), or Kenwood's TH-D72A (Figure 3.7, page 36), there are also units such as Yaesu's FT-65R (Figure 3.8, page 37), that costs about $85 and offers vastly improved features, durability, audio quality, and ease of use over the Baofeng.

Figure 3.6: Yaesu FT-2DR

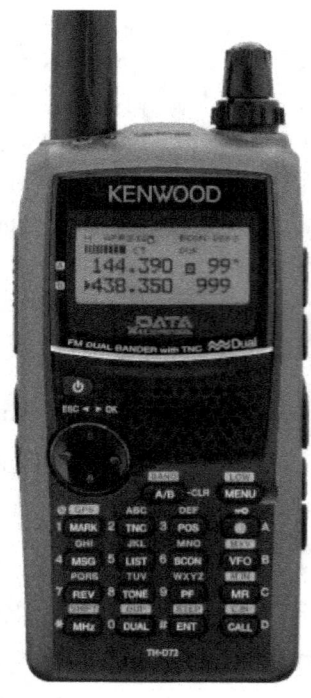

Figure 3.7: Kenwood TH-D72A

Figure 3.8: Yaesu FT-65R

Antennas for Handhelds

You can connect your handheld transceiver to any antenna designed to work on the frequency on which you wish to transmit.

There are rare exceptions, but in general, any quarter-wave 2-meter band antenna will work as a 5/8 wave or half-wave (depending on the design) 70-centimeter band antenna – that's the main reason dual-band radios are 2-meter/70-cm radios rather than, say, 2-meter and 6-meter radios.

The obvious choice for a handheld is the "rubber duckie" antenna that almost certainly came with the radio. Those antennas were developed in the 1960's to replace the long steel whip antennas in use at the time. Rubber duckie antennas are springs encased in some sort of flexible case. It's a lot easier to carry around an eight-inch antenna than a 19-inch antenna! Of course, there's a price to be paid, and that price is efficiency.

You can buy upgraded antennas for your handheld that are still intended for walkaround use. The one I seem to see the most is about 15½ inches long. It has the same electrical length as the coiled-up rubber duckie, but by increasing its physical length, we get more efficiency and gain. Better for transmitting and receiving, not so great for bushwhacking your way through thick underbrush on a hike, or working in a crowd, where you're at more danger of poking someone with your antenna. One more lesson in "antennas are always about compromises."

Figure 3.9: Dual-band J-Pole Antenna (Courtesy of "J-Pole John" Bole, K9JEB https://k9jeb.com/JPole.aspx)

Once we give up walkaround capability and start treating the handheld as simply a small, portable transceiver, we open up lots of possibilities.

Most of us have at least a distant familiarity with the common j-pole antenna. Figure 3.9 shows a dual-band j-pole designed for permanent mounting.

That's not very portable; at least not "backpack" portable. Its electronic cousin, known as a "slim-Jim" antenna, can be.

Slim-Jims don't have to be constructed of rigid tubing – especially not when they're used with handhelds, since they only have to handle 5 to 8 watts of power. This one is made of a length of transmission line known as ladder line.

You can make a slim-Jim out of "window line", aka "ladder line" transmission line, roll it up, and toss it in your backpack. Throw in a compact fishing pole, too, with a heavy round weight on the end of the line. Once you get to your destination, cast the weight over a handy tree limb, use the fishing line to pull up some paracord, then hoist your wire j-pole and feed line as high as you can get it. Who needs a tower when Mother Nature provides them all over the landscape? You can see a very portable dual-band slim-Jim in Figure 3.10.

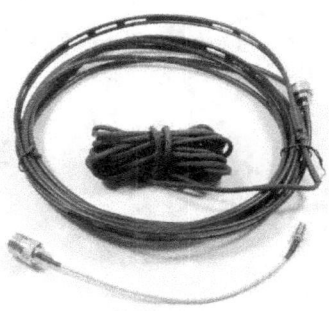

Figure 3.10: Roll-up Slim-Jim Antenna Kit by The Survival Antenna

What if you're in the middle of the Bonneville Salt Flats, though, with not a tree in sight? You might have some interest in something like a Buddipole™. A Buddipole is sort of an antenna Tinkertoy set. You can use the parts to create antennas for anything from 80-meters through 2-meters. The portable tripod tower will put your antenna up to 11 feet or so above the ground; 18 feet with their optional larger tripod. Figure 3.11 shows one down at ground level, being set up as a 2-meter vertical dipole.

Figure 3.11: Buddipole Portable Multi-band Antenna

Chapter 4

Dual-band Mobile Stations

Installing a dual-band mobile station in your vehicle gives you, in my opinion, a more versatile and usually a more often used station than relying solely on a handheld transceiver.

Mobile dual-band radios have, typically, output powers ranging from 25 to 75 watts; lots more than handhelds, which are usually 5 watts. (A few handhelds claim 8 watts, though testing has shown some of those claims are somewhat optimistic.) Even a magnetic mount quarter-wave dual-band antenna on your roof is far superior to a rubber duckie antenna inside your car, so mobile stations start off with some huge advantages over handhelds.

By way of comparison, hitting our club repeater while standing in our driveway with a handheld is a very hit and miss proposition involving standing in just the right spot and holding the radio at just the right angle. Some days the magic works, some days it doesn't. Believe it or not, just having the trees grow leaves in the spring is often enough to bounce us out of the game with a handheld. From the same spot, sitting in our truck with our dual-band mobile, we sound like we're "right next door." In fact, we dial the power back to 25 watts – the full 50 watts is just overkill.

With a mobile unit, drive time becomes ham radio time. In many states, thanks to a lot of lobbying by local ham clubs and the ARRL, it's legal for licensed hams to operate a mobile radio while driving. Obviously, be safe, check your state's laws, don't abuse the privilege, and carry a copy of your license, just in case. Since most mobile radios have a scan function, you don't need to be constantly tending to the radio while you're rolling along at 60 or 70 mph.

If you're up against zoning or homeowners' association rules that forbid an-

Figure 4.1: Yaesu FTM-400DR Mobile Transceiver

tennas, or just have a home where it is physically impossible to put up an antenna, a mobile station might become your main ham shack. It's easy enough to turn it into a "shack with a view," too! One ham I know spends most evenings enjoying the sunset from a beautiful local park near his home while making contacts with his mobile rig. Others go out on weekends to "activate" locations for Summits on the Air, Islands on the Air, Parks on the Air, and similar activities. The world is your ham shack when you're mobile.

Dual-band Mobile Transceivers

Dual-band mobile radios range from very basic models, priced around $100, to very advanced models such as the Yaesu FTM-400DR pictured in Figure 4.1 that includes APRS, built-in GPS, digital voice capability, and what can be a dizzying range of advanced options. Given their capabilities, even the most sophisticated dual-band radios are surprisingly affordable. The Yaesu's list price is currently around $580, and I've seen them on sale for as little as $450. My mobile unit is a Kenwood TM-D710GA, with a list price of $590. (We won that one at a hamfest. Thanks, Seapac!)

Both the Yaesu and Kenwood have detached control heads. In fact, these days almost every radio from the major manufacturers – ICOM, Kenwood, and Yaesu – has a detached control head.

Detached control heads greatly simplify the installation process. While the cab of our Ram 1500 is huge, the space between the front seats is fully occupied by a brawny console and I didn't want to cut any holes in the console. The solution

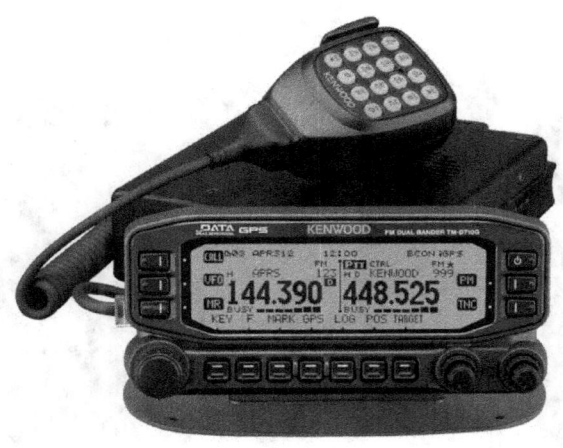

Figure 4.2: Kenwood TM-D710GA

was a detachable head radio on a gooseneck mount that attaches to a seat bolt. We think it makes a nice, tidy install and it leaves plenty of room over there for KC7YL. You can see the control head in Figure 4.3

The actual radio for that unit – the box the control head is controlling – is tucked under the driver's seat. It's a small square box, less than six inches on a side, and maybe an inch-and-a-half thick. There's a control cable that runs under some carpet and between the seats, then is cable tied to the gooseneck.

Detachable control head radios start at around $230 for the Wouxun (woo-SHUN) KG-UV920P, shown in Figure 4.4.

I've never tested a Wouxun radio. My impression of the brand is that they position themselves (and price their products) as a step up from Baofeng. Any time you are pondering an equipment purchase, it's a good idea to read the reviews on eham.net.

https://www.eham.net/reviews/

The reviews on eham.net for this Wouxun radio are not great – but they are much more positive at a well-known US dealer, BuyTwoWayRadios.com.

https://www.buytwowayradios.com/wouxun-kg-uv920p-a.html

Even those reviews seem to indicate some quality control problems with that radio. If you need a detachable head radio, and you can swing it, it is worth it to go with a major brand like ICOM, Kenwood, or Yaesu, in my opinion.

If you can live without the detachable head, prices for 40 to 50 watt dual-band mobiles start around $140 for a QYT KT-980 – which, at least from the outside, looks for all the world like a Baofeng/BTech UV-50X2, which sells for around

Figure 4.3: AF7KB & KC7YL's Kenwood Install

Figure 4.4: Wouxun KG-UV920P

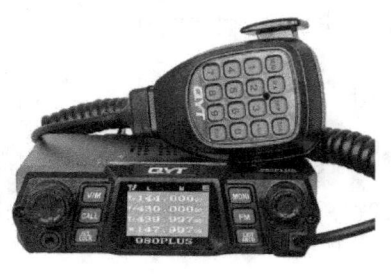

Figure 4.5: QYT KT-980 (left) & BTECH UV-50X2 (right)

$180. (For all I know they're all the same company.) You can see those radios in Figure 4.5.

If you can live with fewer watts, there are units like the QYT KT-8900D, with a claimed output of 25 watts, for $80. Just be aware that most of the Baofeng, QYT, TYT, and other non-ICOM/Kenwood/Yaesu radios are going to be subject to the same manual programming woes as the infamous UV-5R. Before you buy, I'd advise you to make sure CHIRP and/or RT Systems can be used to program the radio.

Before you commit to the radio, take time to consider how you are going to mount it in your vehicle. That BTECH UV-50X2 measures 5.7 in (w) x 1.85 in (h) x 7.5 (d) in and is intended to be mounted with an included mounting bracket. Those brackets are a simple way to mount a radio *if* you have the space. Before you commit, ask yourself: If you mount that under your dashboard, will your knees be bashing into it? The QYT KT-980 measures 4.96 in (w) x 1.85 in (h) x 4.05 in (d) – a more compact package.

Installing Your Dual-band Mobile Radio

Before we get started, please understand this; my skill level with mobile installs, as well as any other craft-related projects in this book is strictly "guy who has worked on some cars and fixed some stuff around the house." I've never made my living building or installing anything. There probably are some far better ways of doing things than I have managed to discover, but I offer you what has worked for me. As always, see to your safety first.

Planning and executing the installation of a mobile radio takes considerable thought and a lot of "measure twice, cut once." The good news is there are components and tools that can make the whole job a lot easier than you might think if you have never taken on such a project.

There are a few parameters to keep in mind:

- Safety. Safety, safety, safety. One idea that seems to occur to a lot of newbies is, "I'll just put it on the dashboard/console/steering wheel/whatever with some double-sided tape/sticky velcro. This is a bad, bad idea. Neither of those methods of securing something is designed to survive a crash, but never mind the crash – back in my early truck driver days, before I had any sense, I'd often mount my microphone clip with some double-sided sticky tape and the clips would regularly fall off just from the vibration of travel. That would drop the microphone to the floor of the cab. Guess what scared me half to death one day by getting stuck under my accelerator pedal? (Modern trucks with a full load don't ride radically different from your four-wheeler, so this wasn't "just something that happens in trucks.") You really don't want loose objects rattling around your space while you're driving, much less hitting you in the head in an accident.

  Just as bad is having a firmly mounted object block your vision or your access to critical controls, so remember to leave room for your knees or shins to clear the radio as you work the pedals. In most states it is illegal to have an object in the "driver's line-of-sight." Anything that blocks even part of the view of the road can get you a ticket or, worse, cause an accident or compromise your legal position in the event of an accident. (An object that blocks your view of the hood and only the hood is, apparently, fine in most states.)

- Airbags. More safety. I've seen several pictures of detachable control heads affixed to the center of the steering wheel. At first glance, this looks like a good idea – the control head is visible and easily reached. Unfortunately, if the airbag in that steering wheel deploys, the control head becomes a 200-mph projectile aimed straight at the driver's chest. You don't want to be the star of that movie. The same principle applies to any location in the cockpit with an airbag behind it.

- Electrical safety. You'll be working with 12-volt electricity, so there's little danger of receiving a serious electrical shock, but there is some danger of burns from arcs and of setting your vehicle on fire. Assuming this is not your objective, you can avoid this by using the proper gauge wire, carefully routing the power wires for your radio, protecting the wires from abrasion and pinching, and installing proper fuses on *both* legs, positive and negative, as near the battery as possible. Before you connect the power wires to the battery – and that is where you want to connect them, to minimize interference from the car's electrical system – and to the radio, put your ohm meter on those leads and be sure you're not hooking up a short circuit you inadvertently created in the course of the installation.

- Antenna placement. As with just about any antenna installation, height is might, so higher on the vehicle can make a big difference in your coverage. Of course, that makes the roof the most desirable location. It's not just higher, it's also not being blocked by the metal of the passenger compartment. Roof mounts are not always practical, though. For one thing, there are those pesky low entryways to parking garages and such to consider. The sound of your $119.00 antenna getting snapped off or bent by one of the bars they put over some WalMart entrances to keep out the big rigs is an expensive bit of music, so put some thought into your antenna placement.

- Antenna grounding. There are some mobile antennas that advertise that they don't need a ground plane, but even those work better when they have a solid electrical connection to the body. Most require it. Without that solid ground plane, those antennas that require one will have a sky-high SWR (Standing Wave Ratio), possibly high enough to cause your radio's self-protection circuit to kick in, effectively shutting off the transmitter. We'll cover some mounts for various locations on vehicles that accomplish this. They also let you mount an antenna without drilling holes in the body – something most people want to avoid doing to their cars.

Professional Installation

One way to approach getting a radio into a vehicle is by having a professional installer do the job.

You might think, "Well, dandy. I'll call up my local car stereo installation shop and they can handle it." Unfortunately, they probably won't. Friends who have tried this route have been mostly turned down. Car stereo installers install car stereos, not ham radios, which are, it seems, very foreign things to them. To be fair, I imagine their insurance would not cover a ham radio installation, either. If you think about it, for the most part they're not "installers" they are more like "replacers."

An approach that almost worked for at least one friend was to pay a local stereo installer to do what I consider the "heavy lifting" of the job; getting power and the feedline from outside the passenger compartment to the radio location. That scheme went off the rails, though, because, despite written orders to the contrary, they did what car stereo installers often do; they took a power feed from inside the cockpit, from an accessory fuse location. That works okay for a low-grade stereo installation, not well for a high-grade stereo, and even worse for an amateur radio. Every major manufacturer has the same recommendation for connecting power to the radio; attach the power wires directly to the battery. We're not trying to hear the strong signal of the local 100,000-watt FM station on our radio, we're trying to pick up relatively faint signals. Our receivers are orders of magnitude more sensitive than commercial broadcast receivers. When it comes to intelligibility, every little bit of extra noise is our enemy, and connecting directly to the battery helps reduce the noise brought into the radio on the power wires.

There are probably some professionals near you who will do the job, and most likely do an amazingly good job at it. They just don't happen to be car stereo installers. They're whomever your local first responders use to install the communications gear in their vehicles.

You'll need to find your way to the first responders' radio technician, but that shouldn't take more than a couple of phone calls. Our club has a member who is an ex-policeman and went this route. He is only recently licensed but has a setup in his SUV that left me jealous! It was not an inexpensive way to go, and the vehicle no longer even resembles a factory stock vehicle inside, but wow! It honestly looks something like Figure 4.6, only without the laptop computer.

Let's assume your needs are at least a bit more modest and take a look at how a do-it-yourself installation might go.

Doing it Yourself

If this is your first mobile install, you can count on it taking the better part of a day, and maybe even a day-and-a-half.

There are, no doubt, many ways to approach the job. Here's a suggested order of operations.

1. Get all the parts.

2. Gather your tools.

Figure 4.6: Police Car Interior

3. Run power wires from the battery location to the transceiver location. (Do not connect power yet.)

4. Install fuses in power wires. Follow manufacturer's recommendations for fuse value. Power wires with fuses pre-installed are often included with high-end radios; count on needing to splice in some extra wire to reach from the battery to the radio.

5. Install correct power connector to the radio end of the power wires. These are also usually included with high-end radios.

6. Install the antenna mount (if any) and antenna.

7. Run feed line from antenna to location of transceiver.

8. Install radio in vehicle.

9. Connect antenna and power wires to radio.

10. Commence playing radio!

Let's go over just what "get all the parts" means.

Unless you are fortunate enough to live in a city with a good ham radio supply store, you'll be doing some internet research to find all the parts. Any of these listed on the next page is a good starting place. I'm just listing them in alphabetical order, not in order of preference.

For radios:
BaofengTech (Baofeng radios) https://baofengtech.com/
BuyTwoWayRadios.com http://buytwowayradios.com
DX Engineering (major brands) http://dxengineering.com
GigaParts (major brands) http://gigaparts.com
Ham Radio Outlet (major brands) http://hamradio.com

For antennas, mounts, and other accessories,
DX Engineering http://dxengineering.com
GigaParts http://gigaparts.com
Ham Radio Outlet http://hamradio.com

Feel free to call any of them for advice before you place your order with them if you're not sure about what goes with what. I think it is safe to say they would rather get you straightened out up front than process a return from you later.

As someone who makes his living selling stuff on Amazon, I'm rather partial to that company; but in this case, my advice is to stay away from Amazon for ham radio gear. Aside from the fact that there is no expert advice available there, you'll usually save money by dealing with one of the ham radio specialty stores. That $9.99 antenna you found on Amazon is no bargain if it has such a high SWR that it fries the final amplifier of your $200 or $600 radio.

All the Parts

The radio. We've covered a lot of possible choices for this.

The antenna. Your choice of antenna will be largely dictated by where and how you want to mount it on your vehicle.

When you go antenna shopping, you'll find lots of choices for mobile antennas – maybe too many! The truth is, there aren't any "miracle antennas" out there; there's only so much a more-or-less straight piece of wire can do in terms of gain and bandwidth, after all.

That last 0.5 dB of gain really isn't the most important parameter when it comes to mobile antennas. Think about what's going to work physically on your vehicle and in your surroundings, then worry about electronic performance and durability. Put simply, how tall is the antenna and where are you going to put it on your vehicle? Consider, too, how easily you'll be able to transfer the antenna to a new vehicle and how important that is to you.

The simplest way to go is a magnetic mount antenna, like the one in Figure 4.7. You can also purchase magnetic mounts without antennas, then add an antenna of your choice. You can find those, and all these other parts, at most of the major online ham equipment dealers.

These typically require a ground plane, and they create one by inductively coupling to the surface on which they're mounted. That won't result in as efficient an electrical connection as a direct connection to bare metal, but that's the price of convenience. The nice thing about mag mounts is that they're so easily removable. Headed for a low garage entrance? Just pull over, pop off the antenna, and toss it in the car.

If you want a slightly more sophisticated installation than "just run the feed line through a door," Comet and Diamond both make window pass-throughs like the one shown in Figure 4.8.

The PL-259 plug on the feed line from the radio attaches to one of those SO-239 sockets, the thin ribbon goes through the window, then the PL-259 on the end

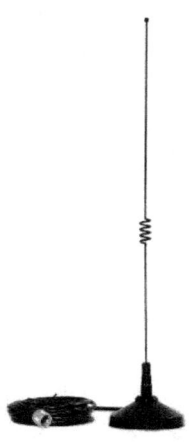

Figure 4.7: Magnetic Mount Antenna

of the feed line from the antenna attaches to the other SO-239. That one handles a maximum power of 60 watts VHF, 40 watts UHF.

Antennas without attached mounts are available for prices ranging from tens of dollars to hundreds of dollars. My choice was a Comet SBB-5NMO. That was based a lot of practical and cosmetic considerations; it's 38 inches tall and does not require a ground plane. I knew I wanted to fender-mount the antenna and I know that doesn't create the greatest ground plane. I didn't want it towering over the cab of my truck. It is black; it matches the truck! Okay, I'm more than a little truck proud, I confess it. I didn't want to create something like the "porcupine car" in Figure 4.9 (page 54).

(That car was well known at various hamfests in its day, and the owner had a great sense of humor about it. He even ran a web site devoted to the Little Red Porcupine Car. I understand he has since passed away, but you'll find something like that at most hamfests. In fact, that's how we know we've found the right place for whatever hamfest we're attending! "Oh, here we go – a porcupine car. This is the place!")

The antenna mount. You'll have a choice to make about what sort of connector the antenna has. Some come with SO-239's, which are the socket into which a PL-259 plugs. Others come with what are called "NMO" connectors. NMO stands for "New Motorola."

You might recall from your Technician studies that PL-259's are not waterproof. This is not good if the connector will be exposed to weather. NMO's are

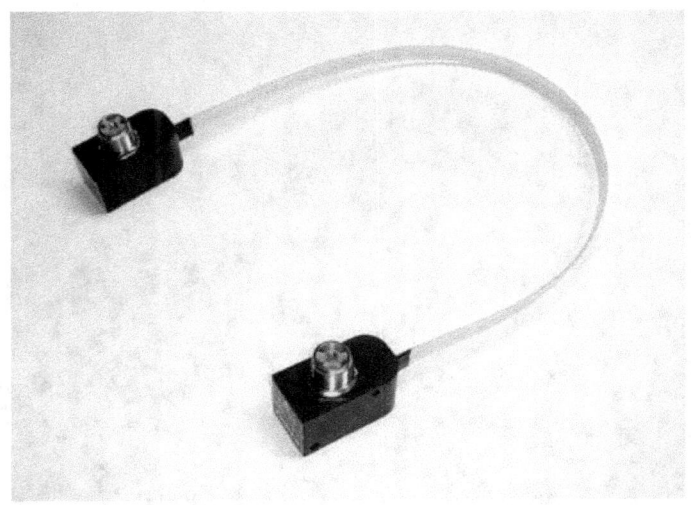

Figure 4.8: Comet CTC-50M "Window Gap Jumper"

waterproof. They're also made to mount through a hole in the chassis. Figure 4.10 shows a through-hole NMO mount with feedline attached.

The first time I went to install one of those, I ended up calling the dealer because I could not figure out how the thing was supposed to work. They patiently explained that the outer ring – where you see a couple of flat sides – unscrews. You make a ¾" hole, unscrew that ring, stick what's left through the hole, and screw the ring on. Tighten it down, and – assuming it's in contact with some bare metal -- you've made your solid connection to ground and created a spot to screw on the antenna.

No holes are necessary, though. Figure 4.11 shows the mount I used on my truck. It's made by Comet and designed specifically for my Ram, but they make them for other vehicles as well. You unbolt a fender bolt under the hood, stick the bolt through a small hole in the mount, reinstall the bolt, and you are done installing the mount. An NMO mount like the one in Figure 4.10 goes through the big hole in the mount. Tighten it down, screw on the antenna, snake the feed line into the passenger compartment, and your antenna is mounted.

Keen-eyed observers will note that the color of that mount is close to that of the truck. It did not come to me that way; out of the box, it's just bare stainless steel. I painted it but masked off the undersides of the pieces where the antenna and the existing bolt install. That way, there's electrical continuity from the antenna base to the entire truck body. That's something you can (and should) check

Figure 4.9: The Famous Little Red Porcupine Car

with your ohmmeter.

Another antenna mount style is what I'll call the "lip style."

The one shown in Figure 4.12 is designed to clip on the forward part of your trunk lid. It's available from MFJ Enterprises, one of the major ham suppliers.

http://mfjenterprises.com

On most mounts like this, pointed set screws underneath pierce the paint on the underside of the lid and provide an electrical connection. The knob is to loosen it if you need to tilt the antenna for some extra clearance somewhere. There are similar designs available to mount to rear door lips.

A feed line: Many NMO mounts come with feed lines attached. In fact, many come with a length of very thin feed line to make it to the other side of a hood, door, or trunk lid, then a length of thicker – and lower loss – feed line to run to the radio. That's usually RG-8X, which is still relatively thin and very flexible; two desirable characteristics in mobile installation. Those multi-diameter feed lines are very convenient, at least for your first installation, so that's what I'd recommend. It's what I used on our Ram.

It's a little tough to see in Figure 4.13 with all the black-on-black, but if you look closely you can see the feed line sneaking into the cab through the door. I could have gone under the hood with that thin part of the feed line, then through

Figure 4.10: NMO Through-hole Mount & Connector

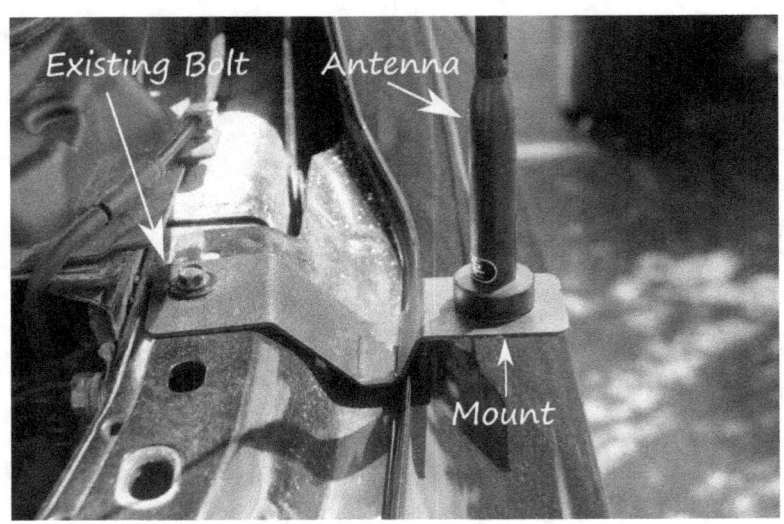

Figure 4.11: Comet Fender-style Antenna Mount

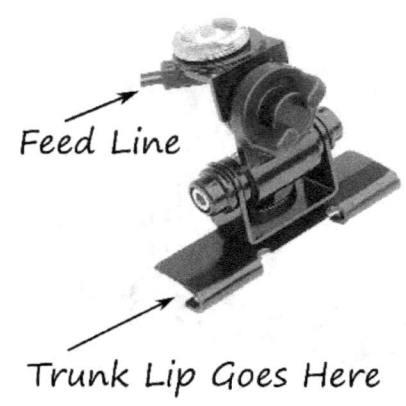

Figure 4.12: MFJ-348M Trunk Lip Mount

the firewall into the cabin, but I wanted to keep the feed line out of the engine bay. Modern engine bays are electrical noise nightmares. I honestly don't know if it made any difference, but we have no noise problem, so something worked!

If you opt for a separate mount and feed line, I'd still point you toward RG-8X. Mobile installation feed line runs are short enough that any difference in performance between RG-8X and stiffer, wider diameter coaxial cable is going to be negligible. One way or another, you're going to have to get that feed line to go where it needs to go in your passenger compartment, and that's much easier with very flexible coax.

Power Wires: It takes two, one positive and one negative. The positive one should be red, the negative (ground) black. (The electricity doesn't know the difference, but using the standard wire colors will help prevent you or some service technician from having a disaster.) As I noted, you'll probably get some of these with your radio, and they'll probably be too short. They usually come with battery connectors and fuses installed on one end and the power connector for the radio installed on the other. If you need to lengthen the wires, you definitely want to save those ends!

Depending on your radio's specifications and the length of the run to the battery, you'll need either 12-gauge (awg) or 14-gauge stranded wire. It's often marketed as "automotive hookup wire" and I've found the best place to get it is usually an auto parts store, where they stock short lengths. Many hardware stores want to sell you a 100-foot roll. That's probably about 90 feet more than you need.

If the wires that came with your radio are 14-gauge, it's perfectly all right to splice in some thicker, 12-gauge wire. Doing the opposite, splicing 14-gauge wire

Figure 4.13: Thin Feed Line for Door Pass-through

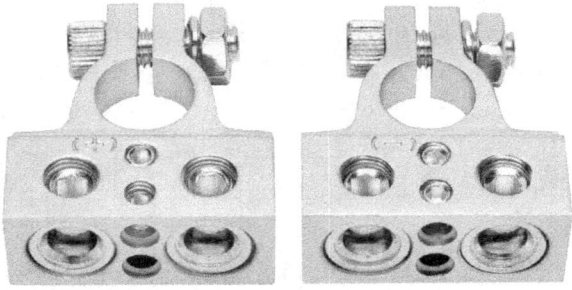

Figure 4.14: InstallGear Battery Terminals

between lengths of 12-gauge is absolutely not all right.

Power Connectors: You'll need connectors for the power wires at the battery end and at the radio end. With luck, those came with your radio. If not, you might consider getting some battery connectors like those shown in Figure 4.14, which make for a really nice, tidy connection, though they may be a bit more than is necessary.

I found the battery terminals in Figure 4.14 on amazon.com, Each of the holes you see accommodates a wire, with gauges ranging from 0 through 10 awg.

Depending on the installation, you may also need some connectors in the middle of the wires to splice in some additional length. Those are called "butt connectors" or "butt connectors." You install them by stripping about a half-inch of

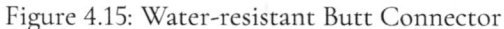

Figure 4.15: Water-resistant Butt Connector

insulation off the end of the wire and crimping the connector onto the bare wire with a special (inexpensive) crimping tool. There are different size connectors for different wire gauges, so be sure you get the ones that match your wire. They come in moisture resistant and non-moisture resistant varieties. The moisture resistant ones – which you want – are translucent, while the non-resistant types are solid colors. After you crimp both ends of a moisture-resistant connector with a crimping tool, you hold a flame under it and the ends shrink down to form a protective seal around the connection.

If you've never worked a crimp connector, I suggest you get your local Ace Hardware electrical person to demonstrate for you. It's really very easy.

Do not be tempted to go all "belt *and* suspenders" and solder a crimp connector. That actually weakens the connection and it will probably break prematurely. Besides, it will completely ruin the water-resistant seal. Solder has no place in vehicle wiring because it is subjected to constant vibration. Aside from the printed circuit boards, there are almost no solder connections in any NASA vehicle, manned or unmanned. I'm thinking NASA probably knows what they're doing in this area!

While you're at the hardware store, pick up a length of what is called **split wire loom**. This is a protective plastic shell you fit over the power wires. It looks much better in your engine bay and passenger compartment than raw wires, but its real purpose is to protect those wires from abrasion which leads to short circuits which lead to The Blues. It comes in various sizes, but half-inch diameter wire loom should be sufficient. In the same department, pick up a pack of **self-locking cable ties**. I like the black ones for vehicles, purely for cosmetic reasons.

Tools

Perhaps surprisingly, no hard-to-find specialty tools are absolutely necessary for this job, though there are a couple that can make it somewhat easier. Almost all of the ones you'll really need are either in your toolbox already or easily available at your local hardware store. You'll want at least:

- A small assortment of wrenches; metric or SAE, whichever fits your vehicle.

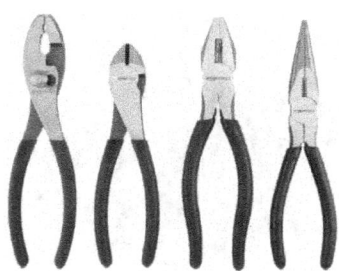

Figure 4.16: Pliers: Slip-joint, Diagonal, Lineman's, Long-nose

Don't assume that because you have an American vehicle it isn't metric; most of the fasteners on my Ram are metric.

- Pliers. Every tool kit needs a set of pliers. The set in Figure 4.16 is from Amazon Basics and costs $15.

Slip-joint pliers are handy if you need to hold onto a nut while you're tightening its bolt. Diagonals are for cutting wire. You probably won't need the lineman's pliers for this job, but they're great for cutting and forming heavy solid-core wire. You also probably won't "officially" need the long-nose pliers, they're for forming small loops at the ends of wires and you are unlikely to need to do that on this job. I always keep a pair close at hand, though; I'm very talented at dropping small parts where they can't be recovered with my fingers, and vehicles present endless opportunities for me to demonstrate my talents.

- A few screwdrivers, both Phillips and flathead.

- If you have a basic set of 3/8" drive sockets and ratchets, like the set in Figure 4.17, those can be useful, especially if you plan to remove any seat bolts – it's hard to get a wrench on to some of those.

- A crimping tool, Figure 4.18.

- A wire stripper. The crimping tool in Figure 4.18 is a "combination stripper, crimper, and wire cutter." My experience is that the "stripper" part of those combination tools is absolute junk, as is the wire cutting part, but maybe that one's great and I've only experienced bad ones; or, I just lack talent. Either way, I far prefer the type of wire stripper shown in Figure 4.19, and so do most of the pros I know.

Figure 4.17: Socket Set

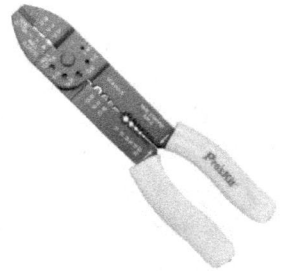

Figure 4.18: Crimping Tool

Figure 4.19: Wire Strippers

Figure 4.20: Trim & Molding Tool Set

To use those, you place the wire in the jaws, running from left to right, leaving the part you want stripped hanging out to the right. You squeeze the handles, one side grips the wire, the other nips into the insulation with a sharp blade, and in one motion your wire is stripped perfectly.

- Trim removal tools, Figure 4.20. These are made of plastic and they're for popping off pieces of interior trim, if you're planning to hide some wires behind trim pieces. You can get a set at tool discounters like Harbor Freight for about $10. Use a strip of painter's masking tape to protect pieces you're prying.

- A bright flashlight. You'll be spending a little time working under the dash, and that's tough enough without trying to do it by sense of touch.

- A utility or craft knife. (Or a sharp pocketknife.)

- Safety glasses or goggles. Especially for working under the dash; dust and miscellaneous stuff often falls out of there when you disturb things.

- An ohmmeter or a continuity tester. This is for, among other things, checking to assure your antenna is properly grounded. The little multimeter from Harbor Freight shown in Figure 4.21 costs about $7 and is perfectly adequate for this job. Sometimes they're free with a coupon.

Perhaps even a little better suited for the specific job we have in mind is a continuity tester. You connect the alligator clip to ground, then use the spike to

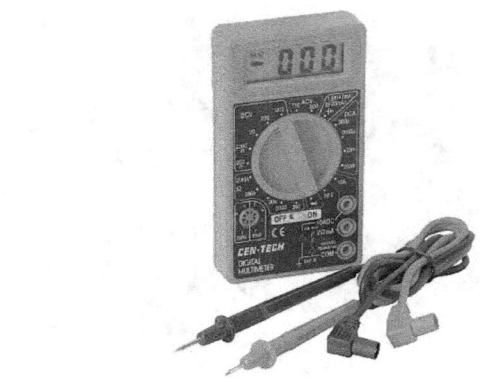

Figure 4.21: Multimeter

Figure 4.22: Continuity Tester

poke at the other end of what you hope is a continuous conductor. It will beep and/or light a light in the handle when it finds continuity. You can usually find these at auto parts stores. They're not expensive – this one is about the same price as that multimeter.

- Access to an antenna analyzer or, at least, an SWR meter that covers the frequencies of the radio.

Let's Install a Radio!

Because of the wide range of vehicles, radios, antennas, and personal preferences, there's no way any book could give you step-by-step specifics for each combination. The basics never change, though. For your particular vehicle, I suggest searching YouTube for "[your vehicle] ham radio installation." Sometimes you'll get very lucky and find a good video of a ham radio installation. More likely, you'll find some bad videos of car stereo or CB installations, but sometimes you'll find that one little clue you need to get yourself unstuck.

Keep safety uppermost in your mind. Doing this job involves some working in awkward positions, especially when you're working under the dash. Work deliberately – there's no rush, and there's no payoff worth getting hurt.

If this sort of work is brand new to you, let me recommend an excellent YouTube video by Eric Hofer, KJ4YZI, who has a channel called HamRadioConcepts. Eric shows a lot of equipment and choices for mobile installations.

https://www.youtube.com/watch?v=UBzENxqQA_w

Plan, Plan, Plan, then Plan Some More

You'll have four major tasks to accomplish; run power lines, mount your antenna, run your feed line, and mount the radio and/or control head.

Your planning will make all the difference in how easily this job goes. (I speak from hard experience, here.)

Don't just "eyeball it" and assume you'll figure it out along the way. That's a good way to end up with either a half-finished job or a pile of extra parts. Get into your vehicle and really figure out where those wires and the radio are going to fit. Don't assume that because my radio unit fit under the driver's seat of my truck that yours will fit under the seat of your Festiva! (Will you be able to hear that radio from under the seat? Will you need a small extension speaker somewhere? Throw a handheld under there, get a ham friend to transmit to you and find out!) Some folks even build mock-ups of the parts out of cardboard to be sure they'll work. You can, after all, get all the measurements from the manufacturers' web sites or their sales reps, assuming you're buying from one of the big three.

When you're planning where to mount the radio, take cooling into account. For an under-the-seat mount, be sure there's some space around the radio. Many have a small fan for cooling, and that needs be able to blow the air somewhere – not crowded up against something that will block the airflow. The radio will only significantly heat while you're transmitting, but if you think about how hot a 50-watt incandescent lightbulb can get, you'll realize that's quite a bit of heat. Similarly, mounting a radio on the dash is going to expose it to the scorching temperatures that can occur in that area in the summertime.

Something to consider: On my Kenwood, and on many (maybe all) the other radios with separate control heads, the microphone plugs into the radio, not the control head. This is probably a better design overall, but I had to buy a microphone extension cord to make that work comfortably – I found mine on e-bay, because Kenwood only sold one bundled with an extension for the control head cable, and I didn't need that.

Power

Let's start by getting some 12-volt power for the radio.

I realize some of the non-major manufacturers helpfully (?) include a plug to go into your vehicle's 12-volt "convenience outlet", which is the euphemism for the cigarette lighter plug. In terms of simple wattage, this would be an adequate (barely ...) power source for most radios. My truck is fairly typical, in that it rates the 12-volt outlets at 13 amps, for a maximum of about 150 watts. That should power even a 50-watt mobile adequately, so long as you don't have your GPS and phone charger plugged in to the same outlet. However, it tends to be noisier power than you'll get by connecting directly to the battery. There's a reason that ICOM, Kenwood, and Yaesu don't include a plug like that with their radios, and it isn't that they can't afford the 25-cent part.

Connecting to the battery will require getting a couple of wires, one red, one black, from the battery terminals through the firewall and through the passenger compartment to the radio location. It might seem, as you consider this, that I foolishly left "electric drill" out of the tool list, but I doubt you'll need to resort to that. There are lots of wires going from the engine bay to the passenger compartment already. You just need to add two more.

On most modern vehicles, you'll find a big bundle of wires going through a soft grommet in the firewall. Today's automakers really go for that "neat and tidy" engine bay look, so it is probably wrapped in black plastic, and might look like some sort of big hose, but look closely and you should find it. That soft grommet is much bigger than the cable going through it and there is plenty of room to do what we need to do harmlessly.

DO NOT POKE, PROD, OR OTHERWISE INJURE THAT BUNDLE OF WIRES. Modern vehicles use a multiplexing system for all those control wires, and if you so much as nick one of those wires you'll probably disable your vehicle and there is no easy fix. It's *very* expensive to have it repaired, and it is not a do-it-yourself job.

All you need to do is cut a small slit in that grommet to the side of that big cable and slip your power wires through. A utility knife, craft knife or even a small sharp pocketknife will do the job, but I use a specialty tool like the one in Figure 4.23.

That's a wire insertion tool, made for exactly the job we're trying to do here,

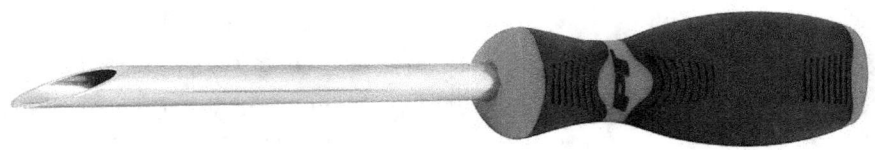

Figure 4.23: Wire Insertion Tool

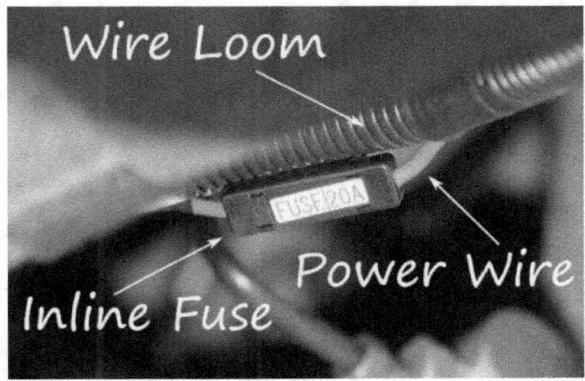

Figure 4.24: Wire Loom and Fuse

and you can find one on Amazon for about $10. The spike you see above is hollow all the way through the handle. You ease the tool through that soft gasket, feed a piece of wire or coat hanger through the tube, pull off the tool, then use the wire or coat hanger to pull your power wires through the gasket from the passenger compartment side. As I said, you can accomplish the same thing without that tool, but I, personally, felt a little safer doing this bit of surgery with that tool.

Once you've gotten the wires through the firewall, and you've figured out your route to your radio, cover the wires with that wire loom. It's just plastic tubing with a split that lets you push the wires into it if you need to, or you can just cut a length and feed it over them from the ends.

While you have the wire loom out, this is probably as good a time as any to cover the power wires under the hood with wire loom as well.

You can see in Figure 4.24 I used a couple of pieces of electrical tape to keep the wire inside the wire loom except where the fuse hangs out.

Next you'll need to route those wires under the dashboard and possibly through the cabin. There's lots of stuff to secure those wires to under your dash, so be generous with the cable ties, take your time, and do a nice neat job under there.

If your radio is not going to be mounted in or very near your dash, you might need to route those wires around or through the passenger compartment. Time to get creative. One popular solution is to hide them under the plastic trim panels that probably surround your door. Those can usually be (gently!) pulled off, and often reveal great hiding places for wires. On my Ram, there's a panel at the bottom of the door opening that pops out and reveals a small but adequate channel at the bottom of the door that holds the wires that run under my driver's seat and power the radio. Other cars have screw-in metal panels covering the same space. You won't know what's in there until you gain access – and this is one place where YouTube can occasionally be a big help, because car manufacturers are pretty good at making everything look seamless when it really isn't.

Antenna

If your only consideration is optimizing your reception and transmission, the best spot for your mobile antenna is dead center of the roof. It's the highest spot with the biggest chunk of metal to serve as a ground plane.

That's easily accomplished with a magnetic mount antenna. That's not the tidiest installation, since the feed line has to trail across your roof and sneak inside through a door. Of course, magnetic mount antennas are useless on aluminum and fiberglass, too.

To achieve a permanent mount, you'll probably need to gain access to the underside of the roof, which means removing the headliner. If you're not an automotive pro, I hope you'll let a professional do the removal and re-installation of that headliner. Getting the headliner down is easy. Getting it back up, looking like new? Not so easy. An automotive upholsterer should be able to handle it, and I imagine most body shops have someone to do the job as well.

Once you have access, you use a three-quarter inch metal hole saw to put a hole in your roof, screw on your NMO connector, screw on the antenna, and you're ready to run the feed line.

There is a way to avoid removing the headliner, but it requires a $60 tool from Laird/Antenex, shown in Figure 4.25. It's a hole saw with a depth stop – it cuts just deep enough to get through the sheet metal of your roof but doesn't cut the headliner. Then you can snake the feed line between the headliner and the roof, get it over to the side, and just remove enough headliner to get ahold of the end, then pull it to where you want it.

If you, like me, start twitching at the thought of drilling a hole in your vehicle, read on.

Figure 4.25: Antenex NMO Installation Saw

We've already covered hood mounts and lip mounts. Those are both non-invasive mounts. We also covered how to install a hood mount. Installing a lip mount is easy. You slide it into position and (gently!) tighten the set screws until they pierce the paint, which they'll do pretty easily. You do not need to crank them down so tight they dent the metal. Unless your car has a huge gap between the trunk lid and the body, that mount should be secure without the set screws.

If none of these suits your application, there are many pages of different mounts to choose from at the major ham dealers. Search "mobile antenna mount" and you'll see everything from license plate mounts to bumper mounts to mounts that fit into the corner of a pick-up bed. Remember the folks at those stores are hams and they'll be able to suggest the right mount for your vehicle. You don't have to know it all, you just have to know whom to ask – and now you do.

Feed Line

There's a truth you must come to embrace when you are installing mobile radios; that truth is that the manufacturer got wires from places like the taillights and the engine bay from where they start to where they end *somehow*. It may not always be obvious how they did it, but I guarantee that they did. Your job with the feed line is the same as it was with the power wires. Find the way by following the manufacturer's wires!

SUV's may be the easiest in this regard. If you mount a lip mount on your rear hatch, you're practically in the passenger compartment already. After that, it's easy enough to hide the feed line under some trim or even under the carpet, so long as it is in a place where it won't get crushed or pinched.

Coupes and sedans can be a bit more challenging, but if you follow the taillight wires, you'll often find a path into the passenger compartment. This might involve removing the rear seat and it helps to have a helper with that one. Worst case scenario, you drill a small hole in the wall between the back seat and the trunk, fill it with a grommet, and feed your wire through. Don't skip the grommet part, that prevents the feed line from chafing from movement.

Pick-up trucks are probably the most challenging if you're trying to bring a feed line from the rear. From the front you can bring it through the same soft grommet as you used for the power lines or bring it through the door as I did on my truck.

From the rear, you'll probably need to get under the truck to find where they ran the taillight wires. They have to be there somewhere! Since there are only a few types of trucks, and since they are popular targets for CB radio installations as well as ham radio installations, you stand a good chance of finding your answer for this one on YouTube.

Depending on your route into the cabin, you may need to remove the PL-259 from the radio end of the feed line, feed the wire through your opening, then attach a new PL-259. If so, you'll need a PL-259 of the proper size for your feed line – they're not all the same. Right now, in my parts box you'll find some PL-259's that are made for half-inch heliax hard line; those aren't going to work at all for RG-8X.

When you purchase PL-259's, you usually have a choice of "crimp-on and solder" or "screw-on and solder" connectors. Since the crimp-on types require a special tool, you'll probably want to opt for the screw-on type. Somewhere in your ham club is someone who is adept at doing these connectors – if you've not done it before, I strongly recommend you learn from someone in person. This is not a skill that translates easily to print nor even to video. There's some touch and finesse to it. At the very least, order yourself half-a-dozen or so connectors; the first one you do will almost certainly not go well. Just count it as your practice PL-259.

The basic procedure is to properly strip the outer insulation, peel back the shield, then strip the proper length of the core insulation, leaving a length of the core conductor. I like to leave enough to poke about one-quarter inch beyond the end of the pin; I'll trim it off after I solder. That core conductor will slide into the pin of the PL-259. Tighten the rest of the connector, then solder the inner conductor to the inside of the pin.

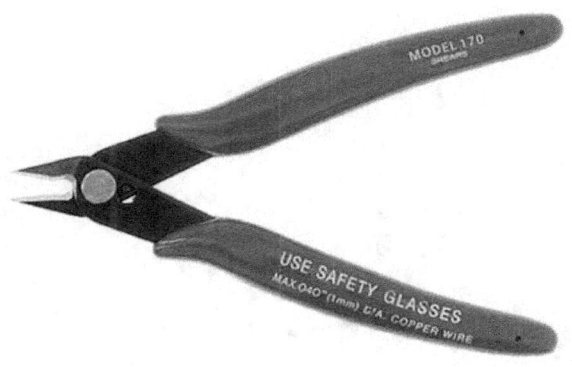

Figure 4.26: Flush Cutters

One tool that is very helpful when working with coax is a pair of "wire nippers" also known as "flush cutters," shown in Figure 4.26. They're great for cutting and trimming that coax shield.

Soldering a big hunk of metal like the pin of a PL-259 takes a powerful soldering iron. I recommend a 100-watt iron. That way the pin will heat quickly, the solder will melt, and the insulator inside the plug won't melt. Using a 60-watt iron puts that inner insulator in a lot of danger because its relatively slow heating gives the heat lots of time to spread throughout the connector.

Congratulations! You've gotten power and the feed line to the radio. Now let's mount the radio and get on the air.

Mounting the Radio/Control Head

At the risk of being dreadfully repetitive, I'll emphasize this again; safety! No loose objects to foul the controls or hit someone in the head.

It's certainly possible, at least with many vehicles, to install your radio in such a way that it looks almost like it came from the manufacturer with a ham radio installed.

That's difficult, though, because most ham radios don't follow the standard sizing of automotive stereo units. I say "most", but I've never found *any* that followed that standard. Besides that, modern car stereos aren't just car stereos – they're often deeply integrated with all sorts of functions in the vehicle.

I imagine you could pull the factory stereo unit and build some sort of shell that would fit in the empty slot to simulate a factory-fresh look. If you have those sorts of skills – well, frankly, I'm not sure why you've read this far in this chapter!

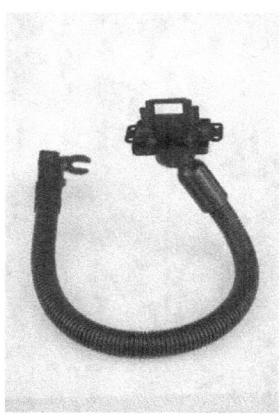

Figure 4.27: Lido Gooseneck Seat Bolt Mount

I certainly don't possess them, and I don't want to pull the music system from my truck, anyway. There are far easier alternatives that are, to my eye, still cosmetically acceptable. Better, they're non-permanent, so when you decide to trade in or sell that vehicle, you pull out the radio and your vehicle is back to normal.

In Figure 4.3 you saw the top of the gooseneck mount that holds the control head for my Kenwood. Figure 4.27 shows the full mount.

Those are available in several different sizes with different mounting systems and adapters for many different types of radios from the folks at Lido Mounts.

https://www.lidomounts.com/hamradiomounts.html

They're simple to attach. You remove one of the bolts that is holding a seat to the floor, then reinstall the bolt through the U-shaped hole you see at the left of the picture of the mount.

It can get even simpler. If you have a center console with a cupholder, Figure 4.28 shows a mount designed to utilize that space.

The big cylinder at the bottom of that mount fits firmly into your cupholder. It adapts to cupholders of various sizes.

Final Steps

You're almost ready to turn that radio on; but hold off just a bit longer. Let's do some final safety checks. Honestly, your rig will probably work just fine the first time you switch it on, but radios are expensive and it's nice to be sure you aren't going to damage yours.

First, just before you hook up the power wires to the battery, use your ohm-

Figure 4.28: Lido Cupholder Mount

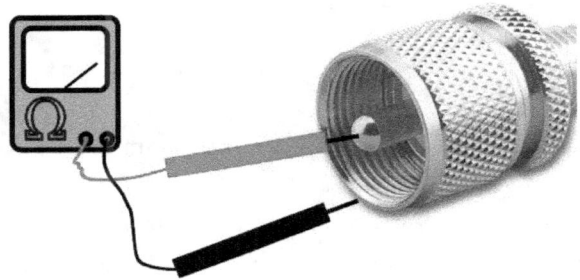

Figure 4.29: Checking for Feed Line Short Circuit

meter or continuity checker to assure you haven't created a short circuit anywhere in the path from the battery to the radio. You'll do that by measuring the resistance from one wire to the other – in other words, you put the black probe on the black wire and the red probe on the red wire, or vice versa. If the meter says "0" or, really, any number at all, you have a short somewhere.

No short circuit? Great! What a relief, right? Attach the red wire to the positive side of the battery, then the black wire to the negative side. Don't attach the power connector to the radio just yet.

While you have that ohmmeter handy, check your feed line for a short circuit, too. You'll measure the resistance between the center pin and the "cap" of the PL-259. Figure 4.29 shows you how to use your multimeter to do that. The resistance should be infinite, or whatever your meter reads when it is set on "ohms" and the probes are not touching anything.

How about checking the SWR of that antenna and feed line?

Ideally, this is when you call your friend who owns an antenna analyzer that will cover the 2-meter and 70-cm bands. Something like the MJF-269C in Figure 4.30.

Figure 4.30: MFJ-269C Antenna Analyzer

That unit costs about $360 and covers all the ham bands from HF through UHF. You connect it to your feed line and antenna, switch it on, select a frequency, and it instantly tells you the SWR at that frequency. It can also tell you the precise resonant frequency of the antenna, so you very quickly know in which direction to adjust the length of your antenna. Ideally, the resonant point of the antenna is right in the middle of the band you're measuring. If you graphed the SWR against frequency, it would look something like Figure 4.31.

That would be an "ideal" SWR curve. In electronics, "ideal" means "never happens in the real world", but you want something resembling that.

If your friend forgot to buy an antenna analyzer, here's a low-budget, but

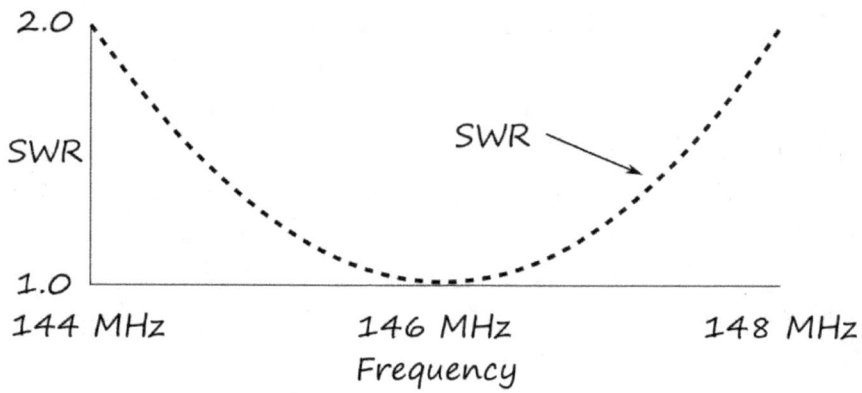

Figure 4.31: (Ideal) SWR Curve

Figure 4.32: SWR Meter/Wattmeter

slightly riskier way to go. I found this VHF/UHF SWR meter/wattmeter in Figure 4.32 on Amazon for about $27.

Here's how you'll use that. You need a short jumper to go from the radio to the meter. There's a picture of a jumper in Figure 4.33. That jumper goes into the SO-239 on the bottom. Your feed line goes in the SO-239 on the top.

Once that's set up, connect the other end of the jumper to the radio and then connect the power connector. Here's my thinking on that order of doing things; I avoid powering up *any* radio without an antenna or dummy load attached. Even though it's unlikely, the radio might have some defect that causes it to immediately go into transmit mode, or an auxiliary piece of equipment could key the transmitter. Transmitters get very unhappy very quickly if they see what is, essentially, an infinite SWR!

Set your radio to its lowest transmit power. Pick an open frequency within the ham bands near the middle of the band you're testing that is *not* your local repeater's and is also not one of the national calling or simplex frequencies, 144.200, 146.520, 446.000, or 432.10 MHz. Set the meter to read SWR. Keep your eye on the needle as you key the microphone for about a half-second – just long enough to see what the SWR is. Is it under 2:1? Close enough for now and probably forever! Key the microphone again and say your call sign and "testing." (Remember, you must ID even if you are just testing.) SWR higher than 2:1? There's at least a minor problem somewhere, but if it's under 5:1 or so, you can work with it, at least in brief bursts of transmission.

Figure 4.33: Coax Jumper

Remember your SWR reading and lower your frequency as much as you can and still stay in the ham bands, off local repeaters, and off the simplex and calling frequencies. Key the mic again and check the reading. Did it go down as a result of lowering the frequency? That's an indication that your antenna is just a little long. There's probably a set screw or two that will let you adjust its length, as shown in Figure 4.34. Try shortening it no more than a half-inch. There's a chart to help you with your diagnosis in Table 4.1.

Figure 4.34: Antenna Length Adjustment Set Screws

If there's no set screw and no other way to adjust the length of the antenna, well – you say to yourself, "It is what it is," and that's pretty much that. If the SWR is very high, 3:1 or worse, read on – you may not have grounded your antenna well.

Remember, what you're working toward is having the minimum SWR be near the middle of the band.

Frequency Relative to Middle of the Band	SWR	Diagnosis
Lower	Goes down	Antenna too long
Lower	Goes up	Antenna too short
Higher	Goes down	Antenna too short
Higher	Goes up	Antenna too long

Table 4.1: Antenna Tuning With an SWR Meter

Try your SWR readings again, starting at that middle frequency. Do not start hacking length off the antenna – we're not there yet! There's little reason to think that a competent antenna manufacturer sent you a mis-tuned antenna.

(Now you know why people buy $300 antenna analyzers!)

Once you think that half of the band is dialed in, be sure to check SWR at the top of the band – you may have overdone the shortening and need to sneak a little length back into it.

SWR just won't get under 2:1 anywhere on the band? Well, sometimes that's just the way it is, due to a variety of factors in mobile installations. Don't freak out; 2:1 isn't that big a deal. It costs you 0.5 dB, about 11% of your signal, so you'll run high power when other people might run medium power. Most likely, though, you didn't create a good connection to ground for your antenna.

Using your ohmmeter, check for continuity between the base of the antenna and the car body. This can involve some athletics in some cases – make it a little easier on yourself by screwing the antenna off that NMO base so you can make an easy connection there with one of the meter probes. There should be nearly zero resistance to a nearby piece of the bodywork. You may need to poke around some to find a bare piece of metal, or even remove a bolt or screw to find some bare metal. This is where that sharp tip on a continuity tester comes in handy.

Obviously, this doesn't apply to magnetic mount antennas. If that's what you're using, is the antenna mounted on a big piece of sheet metal? That's what it wants to see. Years ago, before I knew anything about SWR, I was trying to make a CB mag mount antenna work on a Freightliner. I thought I was very clever to come up with a piece of steel that bolted to one of the mirror mounts to hold the antenna. (This was in the days of "West Coast mirrors", not the new streamlined truck mirrors.) The joke was on me, though – Freightliner bodies are mostly made of plastic. The only ground plane that poor antenna had was the mirror mount.

I never did get the SWR on that thing below 5:1 and I was lucky to get a signal a quarter mile down the road.

Checking for ground continuity also doesn't apply 100% to "no ground plane needed" antennas, but even those work better with a ground plane.

Once the antenna system checks out properly, you're ready to put away the tools and start playing radio.

Chapter 5

Dual-band Fixed Stations

Choosing a Fixed Station Radio

There was a time when a "fixed" or "base" station for VHF/UHF would be a multi-mode transceiver, capable of operating in Upper Sideband, Lower Sideband, FM, and even CW modes on the VHF – and sometimes UHF – bands. It would look something like the ICOM IC-271A you see in Figure 5.1 which was a popular 2-meter base station radio in the mid-1980's. If you found a bargain price on one at a swap meet and you mostly work one 2-meter repeater, it would be a dandy home radio provided you can make your hop with 25 watts.

The nearly complete absence of multi-mode VHF/UHF radios like the one in Figure 5.1 from the marketplace suggests that FM is *the* mode of choice on VHF/UHF for most operators.

Hard-core advanced VHF/UHF operators might have a carefully preserved

Figure 5.1: ICOM IC-271A

radio like the one above, along with an RF power amplifier and a directional antenna on a tower. For most of us, though, dual-band fixed stations use the same radios as our dual-band mobile installations. In some cases, it is literally the same radio shuttling back and forth from home to car and back.

Your VHF/UHF "Shack"

Since any modern dual-band mobile radio is going to be quite compact, selecting the spot in your home for your VHF/UHF ham shack is considerably simpler than doing the same exercise for an HF station, because HF stations typically require more equipment.

Here are some key considerations for your decision.

- Adequate space for a desk or table to hold a (small) radio and power supply.
- Room in front of your operating position for your elbows and a notepad.
- Easy access to an electrical socket.
- Placement on an outside wall simplifies getting the feed line through the wall and up to the antenna.
- A comfortable chair.
- Climate control. You won't want to put a nice radio or yourself in a damp, cold basement.
- A quiet location, both sonically and electromagnetically. You can check the electromagnetic noise level of a potential location with a portable AM radio tuned to a space between stations.

If you have children in the house, having a way to secure your station is a very good thing; after all, you're responsible for that station.

Since mobile radios operate on "12-volt" power, you'll need a power supply. You'll also need an antenna, a feed line, and, if that antenna is permanently installed, a ground system. We'll cover each of those items in turn.

Power Supply

Your power supply needs to be a source of well-regulated, quiet 13.8-volt DC power capable of supplying adequate amperage to run your radio. (13.8 volts is the

Figure 5.2: MFJ-4115 15 Amp Power Supply

real operating voltage of your car's "12-volt" system, because it takes more than 12 volts to charge a 12-volt battery.) Well-regulated means the power supply voltage remains at or at least very near 13.8 volts under all conditions – especially when your radio is transmitting. Quiet means the power supply itself is not a source of RF interference. Adequate amperage will be one of your radio's specifications. As a rule of thumb, take the radio's highest transmit power and double it. For a 50-watt radio, then, you'd want a power supply capable of supplying at least 100 watts at 13.8 volts. To get the amperage, we divide the watts by the volts and that comes out to about 7.25 amps. We need a power supply capable of supplying *at least* 7.25 amps; more is fine. It would be unlikely you'd find a 7.25-amp power supply though; it would probably be a 15-amp power supply, such as the one shown in Figure 5.2 from MFJ Enterprises.

The power supply is an area of your station where it is tempting to shortchange quality. I'd advise against that. There are very inexpensive power supplies available from e-bay and similar sources that advertise themselves as 13.8 volt/30-amp power supplies for around $25. I've never run a radio with one, but I'm deeply suspicious that they would produce a lot of RF interference. Most suppliers advertise those power supplies as being for powering strings of LED lighting, not for powering radios. They are switching power supplies, and unless a switching power supply is designed to be a low-noise power supply (as the one above is) the constant switching of the final amplifier in a switching power supply is a source of very annoying RF interference. That's why the e-bay power supplies cost $25

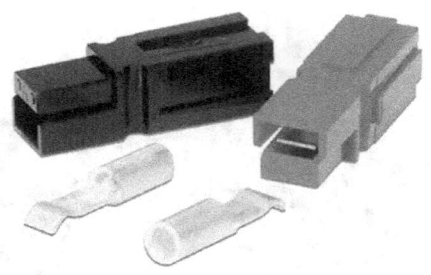

Figure 5.3: Anderson Powerpoles

and the MFJ power supply costs $74.95.

Before you jump on that 15-amp power supply, you might pause and consider whether there is an HF rig in your future. Most modern HF transceivers require minimum 25-amp power supplies, and it isn't terribly expensive to make the jump from 15 amps to 25 amps. If you go for the 25-amp supply, you'll be all set for HF, and you'll only have one power supply box on your desk. We run our fixed rig off a 35-amp Astron power supply. It's built like a battleship and is very, very quiet because it is a linear power supply, not a switching power supply.

As you are hooking up your power supply, you're almost certain to run across the term "Powerpoles." Powerpoles are a family of quick-connect power connectors made by Anderson Power Products©. You can see a picture of Powerpoles, courtesy of Quicksilver Radio, www.qsradio.com, in Figure 5.3. Powerpoles solve a challenge in the ham radio world; there's no power connector that is standard for all radios. In fact, connectors are not even standardized across all the radios made by any single manufacturer. This can make things inconvenient in a home installation if you want to connect different radios to your power supply at different times, but can be disastrous in an emergency communications operation when radios may need to be swapped out quickly due to failure, different capabilities, etc. For this reason, the ARES and RACES folks long ago standardized on Powerpole connectors for all their radios.

Powerpole connects are very easy to operate; just push together or pull apart. They're even easy to install, though they require a special crimper. (Someone in your ham club probably has one.)

The only problem Powerpoles solve is interoperability – they're not necessary unless you a) plan to operate in ARES or RACES (emergency communications) and/or b) want to conveniently switch which radio is connected to your power supply.

Figure 5.4: Cushcraft A270-6S Dual-band Yagi

Antenna

Any of the types of antennas listed in the section on handheld transceivers should work just fine with a mobile/base radio as well, so long as it is capable of handling the power output of the radio.

Let me suggest you go with a j-pole or a commercial vertical for your first dual-band antenna unless you have a special situation that demands a more directional antenna.

The Tram vertical antenna shown in the handheld section should perform well in most situations. There are similar products available from Comet, Diamond, and Cushcraft.

Stepping up to an antenna with significantly higher gain than any of the verticals will most likely mean stepping up to something like a dual-band Yagi, such as the Cushcraft A270-10S, shown in Figure 5.4, which claims 10 dBi of forward gain and a 20 dBi front-to-back ratio on the 2-meter band, and 18 dBi front-to-back on the 70-cm band.

Notice that antenna is mounted for vertical polarization, which is what you want for repeater work. If you decide to pursue some more advanced VHF/UHF modes, you can always turn it to horizontal, which is what you'll want for some of those modes.

Directional antennas are usually paired with antenna rotors, and Cushcraft recommends a heavy-duty rotor for even that antenna. That item will set you back $500 - $600.

I can imagine a situation where you'd neither need nor want a rotor for your directional antenna; that would be if every radio to which you intend to speak lies in one direction from your home, *and* those radios are at a distance that can't quite be managed by a vertical antenna. That would almost certainly mean that either your antenna site or the other radio's antenna site (or both) enjoyed a considerable height advantage over the surrounding terrain.

For most applications, though, your most practical choice of a dual-band antenna is going to be that vertical.

The real key to antenna performance is the quality of the antenna site. Again, height is might, especially on VHF/UHF. The higher your antenna is, the farther it can see, regardless of its design.

One way to gain height is to have the foresight to purchase a home that sits atop a mountain, with as few other mountains around as possible. For most of us, our home's altitude is pretty much a given; it's unlikely that, at this stage of our ham career, we're going to pull up roots and buy a new home based solely on its HAAT – Height Above Average Terrain. That leaves gaining as much height as we can through various structures.

Gaining even a relatively small amount of altitude for the antenna can provide a lot more coverage area. If we assume the Earth is a perfectly even sphere, the distance to the visual horizon for an antenna five feet above the ground is about 2.7 miles. The radio horizon will be a little farther, but not much. If we can raise that antenna just another 15 feet to 20 feet above the ground, the horizon flies out to 5.5 miles, better than double.

Please note this does *not* mean your antenna mounted 20 feet above ground is limited to 5.5 miles of coverage – if that antenna on the other end is also 20 feet up, it's about 11 miles to the horizon. If that receiving antenna happens to be up 1,000 feet on a mountain, we're up to a line-of-sight distance of over 44 miles. Height is might!

How to get that antenna as high as possible? As mentioned in the section on handhelds, you can always hoist a j-pole or slim-jim style antenna high in a handy tree. If you don't have a handy tree, start looking for something else tall on your

Figure 5.5: Gable Mount

property.

For many of us, that search starts and ends at the roof of our house, so we need a way to mount that antenna on a roof.

Antenna Mounts

An antenna mount has two jobs; to hold your antenna securely so it doesn't fly into your neighbor's bedroom window in the next strong wind, and to hold your antenna in the proper position for transmitting and receiving. It should accomplish those jobs with maximum safety for you when you install it and when you service the antenna.

If you have a peaked roofline, you could consider constructing something like the gable mount shown in Figure 5.5. I cobbled that up from a couple of 2 x 6's and 1 x 2's. The mast is a piece of 2-inch steel electrical conduit, and it's held on by some 2-inch U-bolts.

The 2 x 6's are attached to the eaves by 3/8" x 3" lag bolts. Since that mount holds a mast plus vertical dual-band antenna with a combined height of about 12 feet, it needs to be a sturdy mount, because that's quite a bit of leverage. I couldn't find a commercially made gable mount that looked like it was stout enough to do the job.

Figure 5.6: Rohn Non-penetrating Flat Roof Mount

If you have a flat roof, the Rohn company makes the non-penetrating flat roof mount shown in Figure 5.6. As you can see, you weight it down with 16-inch concrete blocks to keep it in place, so you don't have to put any holes in your roof.

They also make similar mounts for pitched roofs. Their mounts come in a variety of strengths (and prices) depending on your needs.

Of course, the ultimate antenna mount is a tower, as tall as possible. That represents a significant investment in time, money, and, in many areas of the country, regulatory inconvenience. While tower building is well beyond the scope of this book, and probably well beyond your present ambitions, I will mention that there are several different types of towers, including roof mount designs, telescoping "crank-up" towers, and tilt-over towers. Some towers require guy wires, others are self-supporting.

Feed Line

While there are other types of feed line, for your first installation, let's keep it simple and use coaxial cable.

Coaxial cable was invented in the late 1870's by Oliver Heaviside. I'll let the inventor explain its purpose and workings in the somewhat flowery language of the time.

When a number of wires run parallel to one another, either suspended or otherwise, any change in the current flowing in one wire causes currents in all the rest by induction, and the effect may be so great as to seriously interfere with the working of telephonic circuits, and to a less degree of ordinary telegraphic circuits also.

My improvements have for object to obtain perfect protection, and to render a circuit completely independent under all circumstances of external inductive influences. For this purpose, I use two insulated conductors for the circuit, and place one of them inside the other; thus, one conductor may be a wire, and the other a tube or sheath, which must also be insulated. When the tube and inner wire are electrically connected at both ends of the line, as through apparatus in the usual manner, the circuit as thus described is completely independent of other circuits, and any number of such circuits, each containing an insulated tube and inner wire, may be laid side by side without any mutual inductive interference, and without interference from other wires worked in ordinary manners.

Not all coaxial cable is created equal. Let's look at the general anatomy of a coaxial cable, shown in Figure 5.7.

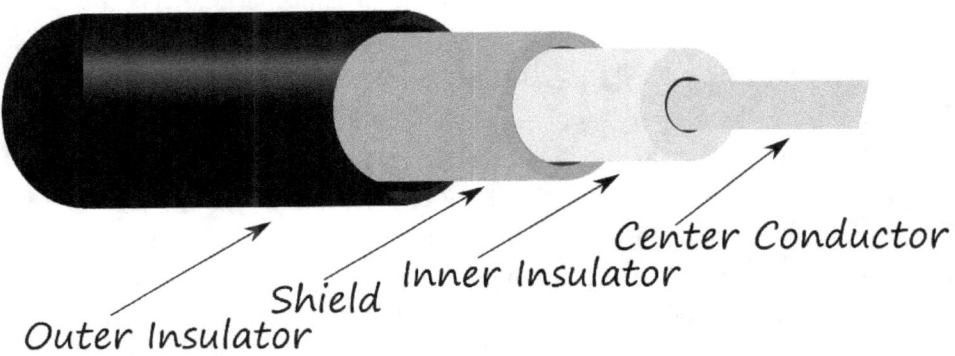

Figure 5.7: Anatomy of a Coaxial Cable

At its simplest, a coaxial cable consists of a center conductor surrounded by an inner insulator (the "dielectric") then a shield, and finally an outer insulator. It's called a coaxial cable because the inner conductor and the shield share a common axis.

Engineers have created many variations on that basic theme. The variations are created by changing the materials of the components and their dimensions.

For example, the center conductor can be solid wire or stranded. The inner insulator can be various kinds of plastic or even air. There can even be more than one shield, for extra protection from noise. As each of those parameters changes, the performance of the cable changes.

Selecting a coaxial cable for a mobile installation is very easy; RG-8X will do the job. Most truck stops sell lengths of RG-58 for the same purpose, and it is acceptable, too.

Once we're no longer installing in a vehicle, a lot more possibilities open up for our feed line.

You'll have some decisions to make; those decisions will guide you directly to the right feed line for you.

There are many different types and sizes of coaxial cable suitable for ham radio use, and none of it comes from a big-box hardware store.

You'll want 50-ohm coaxial cable, not the 75-ohm cable you'll find at the hardware store. You'll also want coaxial cable with adequate power handling capabilities for your maximum transmission power at its maximum duty cycle. (One minute of 100-watt SSB phone is a very different duty cycle from one minute of 100-watt RTTY (Radio-teletype) or SSTV (Slow Scan TV), for example.)

There's no avoiding the fact that coaxial cable losses increase with frequency. That means you need to decide how important maximizing your UHF coverage is to you.

RG-8X was fine for the mobile installation, because for that installation we were talking about a very short run of cable, and the size and flexibility of that cable far outweighed any considerations about loss at the UHF frequencies. However, typical RG-8X loses about 8.1 dB per 100 feet at 440 MHz. That means your 40 watts of output on UHF is a mere 6 watts by the time it gets to the antenna if your RG-8X feed line is 100 feet long – and that's assuming 1:1 SWR, which you almost certainly won't have across the 70 cm band.

If you need a feedline 50 feet or longer to get to your antenna from your operating position, and you want excellent performance in the UHF region, you'll probably want to consider LM-400 or an equivalent cable. If it's *really* important to you to squeeze every watt out the antenna on UHF, you might even consider half-inch hard line coax.

If you're in mountainous country, like where we live, coax loss at high frequencies is not the big consideration it might be in the middle of a flat space like Nebraska or Kansas. Here in the hills and mountains of Washington, elevation is everything. No matter how many watts you have, you're not going to put a signal

Attenuation (dB per 100 feet)						
MHz	30	50	146	150	440	450
RG-174	5.5	6.6	13.0		25.0	
LMR-100A®	3.9	5.1	8.8	8.9	15.6	15.8
RG-58A/U	2.5	4.1	6.1	6.1	10.4	10.6
LMR-200®	1.8	2.3	3.9	4.0	6.9	7.0
RG-59		2.4			7.6	
RG-8X	2.0	2.1	4.5	4.7	8.1	8.6
LMR-240®	1.3	1.7	3.0	3.0	5.2	5.3
LMR-240 Ultra®	1.3	1.7	3.0	3.0	5.2	5.3
RG-8/U FOAM		1.2				
RG-213		1.5	2.8	2.8	5.1	5.1
RG-214	1.2	1.6	2.8	2.8	5.1	5.1
LMR-400®	0.7	0.9	1.5	1.5	2.7	2.7
LMR-400 Ultra®	0.7	0.9	1.5	1.5	2.7	2.7
DRF-400	0.7	0.9	1.5			2.7
Bury-FLEX™		1.1				
Belden RF300 (Direct bury)	1.0	1.3		2.2		3.9

Values indicated are *approximate* and for comparison purposes only.

LMR® is a registered trademark of Times Microwave Systems.

These ratings have been gathered from various sources, including manufacturers' data sheets. In some cases, no data was available from any credible source, so none is reported here. In other cases, values were estimated from published data. Cables with similar – even identical – names but from different manufacturers often have different characteristics.

No recommendation or endorsement is implied by these listings.

Table 5.1: Coaxial Cable Losses

through Mt. Rainier, nor any of the other numerous big piles of rocks that dot our landscape, so overachieving in the area of power is just a waste of electricity. Out in the flats, though, distance transmitted correlates directly with wattage and antenna gain.

The better your coaxial cable's performance at 70 centimeters, the better it will be at 2-meters.

You can take a look at the loss performance of various types of coaxial cable in Table 5.1, on page 87. All figures listed are dB of loss per 100 feet, which is the industry standard.

At some point, the feed line will need to make its way from inside your ham shack to outside the ham shack, then to the antenna. The path that feed line takes will affect your choice of coaxial cable. "Out the window and up to the roof" is one scenario, while "through the wall and underground to the tower" is quite another, and each requires a different sort of coaxial cable.

Coaxial cable can be rated "outdoor use", "indoor use," or "indoor/outdoor use." Within any give cable type, "indoor/outdoor" is usually the most expensive. Cable can also be "direct bury", which means it can be buried without a protective conduit around it.

An "out the window and up to the roof" feed line should be outdoor rated. (Indoor/outdoor would be fine, but why waste money?) The jumper going from your radio to the surge protector should be indoor rated.

Another rating to be aware of is the cable's "plenum" rating. Plenum is a building industry term for the spaces between walls and above and below ceilings and floors. In this application, specifically, it's a space with some sort of ventilation connection to the living space of the building. The space under the floor would usually not be considered plenum space, the space above the ceiling might. Cable that is rated for plenum use is made of materials that, in a fire, produce fumes that are less toxic than those produced by non-plenum rated cable.

A rating that is related to the cable's plenum rating is its "riser" rating. If a cable is "riser" rated, it is acceptable for use in runs that go from floor to floor inside a wall, but not acceptable in plenum spaces.

The likelihood that you need cable that is either plenum or riser rated in an average single-family detached residence ham radio station is low, so unless you need plenum or riser rated cable, there is no point in paying the premium for those ratings. If you have any doubt at all, consult a licensed electrician.

The last feed line consideration is still an important one; how easy is it to work with the coaxial cable? There are a couple of factors that play into this. The first is flexibility. Each coaxial cable can make a certain radius of bend. Bend it any tighter and the cable will be damaged, which will affect your SWR negatively, and perhaps even create a short or open circuit. Some applications demand a very flexible cable, others can tolerate a relatively stiff cable. The second factor is how easy it is to fit connectors on the cable.

At the moment, I have some half-inch "heliax" cable in my back yard that we're preparing to install on a dual-band antenna. While most common amateur radio coaxial cable has a woven copper shield, heliax's shield is solid copper. It looks something like copper plumbing pipe that has been corrugated. The corrugations give it more flexibility than plumbing pipe, but not a lot more. It also has a solid-core conductor, which makes it even stiffer.

As you might imagine, that heliax is some stiff stuff. The bends we'll put in it on the way to the transceiver will have a radius of about a foot or more. Fitting connectors requires a special tool that fits on a power drill and has precision blades to strip the outer insulation, then cut the copper shield and inner insulator to fit a PL-239. That tool retails for about $200, though you can find them for about half that on e-bay. This cable is neither flexible nor easy to fit with connectors! It is, however, very low loss cable, and I'm lucky enough to have a friend who loaned me that fancy tool.

I also have a 100' coil of Messi & Paoloni Airborne 10 (10 mm/.400 in) cable from MFJ Enterprises. That also has a solid center conductor but has a braided shield and very flexible outer insulation, so it is a very flexible cable, considering its diameter. It also has a double layer of outer insulation as well as double shielding, and it is suitable for direct bury – which was important in the application in which it will be used.

We'll go into more detail about how everything connects together, but for purposes of planning your feed line, you need to know that just outside the building where your transceiver is, where the feed line departs toward the antenna, there needs to be a small box known as a "lightning surge protector." You can see one in Figure 5.8. The lightning surge protector has two coax connections on it. One connects to your feed line, the other to a short length of coax, known as a "jumper," that connects to your transceiver. A third connection on the lightning arrestor is for a ground wire that goes to a ground rod. That connects the shield of the coaxial cable to ground. We'll cover grounding in more detail in the next chapter.

The lightning surge protector shown in Figure 5.8 has an SO-239 on the input and a PL-259 on the output. They can also be ordered with two SO-239's, so be sure you order the one that matches your intentions.

There's a schematic of the correct way to connect the lightning/surge protector in Figure 5.9.

The lightning/surge protector is intended to instantly disconnect your equipment from the feed line in the event of a nearby lightning strike – which will induce a very high voltage in your feed line – and route the surge to ground instead of into your equipment. The one above advertises that it is capable of shunting 5,000 amps to ground. That's a lot of current, but it is a drop in the bucket compared to a direct lightning strike, and the specs for that protector specifically state, "does not protect against a direct lightning hit."

Figure 5.8: Lightning Surge Protector - Note ground connection at bottom.

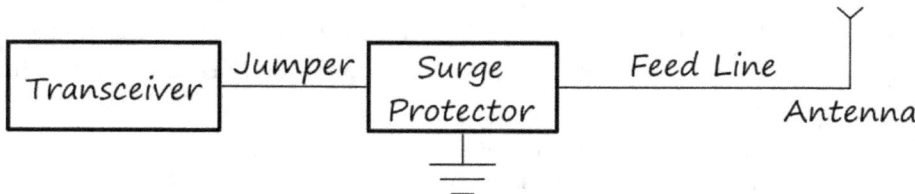

Figure 5.9: Lightning Surge Protector Connection

I mention that because some people have, it seems, misread the name of the device, and think it is a lightning *and* surge protector, but it does not constitute a complete lightning protection system.

Chapter 6

Grounding the Amateur Radio Station

We're going to be delving into the topic of grounding for amateur radio stations.

While this chapter appears in the VHF/UHF section of the book, everything here is equally applicable to HF installations.

Every radio station requires a ground. Sometimes that means an electrical connection to the Earth, sometimes it's an electrical connection to whichever pole of the power source for the radio is, somewhat arbitrarily, designated by the designer as ground.

The first thing I want to tell you about grounding for amateur radio is to believe *nothing* you read on the internet or hear from other hams about grounding. Especially, believe nothing on this topic that you read or watch on the fountains of misinformation called Facebook and YouTube. Don't even believe *me*; I had an expert on grounding review this section in detail, but there isn't space in this book to cover all the aspects of proper grounding.

Here's who to believe: Believe the National Electrical Code. Believe the Motorola document known as *Standards and Guidelines for Communication Sites,* which covers not only grounding but every other technical aspect of building a communications site. Believe H. Ward Silver's ARRL book, *Grounding and Bonding for the Radio Amateur.* If you choose to, go ahead and believe all the internet hooey you want about radios and antennas and such; you'll be ignorant, but harmless. Bad grounding, however, can make you or somebody else dead, and the amount of misinformation available is truly staggering.

There are three distinct but interconnected ground systems in a properly designed and installed station. **None of them are optional.** They are electrical shock prevention grounding, called safety grounding; radio frequency grounding, and; lightning protection grounding.

- **Safety Grounding**. This works the same way the grounds on your household appliances work. If the system is working correctly, any voltage that ends up in an area that could make contact with your body gets carried off, instead, to ground. That round prong on most of your wall plugs is the ground connection. You might notice it is just a bit longer than the other prongs; that's so it makes a connection before any electricity can reach the device being plugged in.

- **RF Grounding**. Of course, there's constant RF energy all around us no matter where we are on planet Earth but sitting at ground zero of even a 100-watt radio station puts us in a field that is orders of magnitude stronger than average. All that RF wants to energize anything metallic in our station. It may even be coming back down our feed line in the form of what are called common-mode currents. If that RF gets inside our transceiver, it creates all sorts of distortion. If it gets on the case of our transceiver, we can get RF zaps every time we touch the radio while it is transmitting. Ouch! Somewhere on the back panel of your transceiver there's probably a place to attach a ground. That ground's job is to conduct RF energy away from your shack. If you have an antenna tuner, that needs to be grounded, too, as does the power supply for your transceiver. The ground conductors should be good, thick ground *braid* or equivalent copper wire, not some skinny little wire. All the equipment grounds get connected to a *ground bus*, which can be as simple as a length of copper pipe with connections made with stainless steel hose clamps or ground rod clamps. From there, one big ground braid or copper wire goes to a *ground rod* driven into the ground as close to your shack as possible. That ground rod needs to be electrically connected, with heavy gauge copper wire, to the house ground rod which should be located near your outdoor electrical service panel.

- **Lightning Protection Grounding.** We'll cover this in more detail later, but lightning protection grounding is an essential part of *any* outdoor antenna system.

Additionally, you may have one more "ground" system, depending on your choice of antenna; that's a **ground radial** or **counterpoise** for your antenna. That system is electrically separate from the other systems and has a completely different function. In this section, we're just talking about safety grounding, RF grounding and lightning protection grounding.

Fundamentals of Grounding

There's nothing electrically magical about dirt. I know they taught you in school that "electricity always wants to go to ground" and that's true, but ground does not necessarily equal the Earth. Connect one terminal of a battery to the dirt in your yard – nothing interesting will happen. No electricity will flow. For the negative pole of that battery, "ground" is the positive pole. For the positive pole, ground is the negative pole.

For your household AC, the "other pole of the battery" is the Earth simply because the power company uses the Earth for an auxiliary "return line" to the generator or the nearest substation. (In a very few old, rural power systems, the ground is the main return line, but that's the exception.) The neutral wire in your electrical outlet is connected to the "neutral bus" in the circuit breaker box, which is, in turn, connected to the "ground bus", and the ground bus is connected to a ground rod – or, heaven help us, maybe a water pipe – outside your house. Then, about every quarter-mile along the electric lines feeding your home, there's another ground rod connected to the ground wire up above, at the top of the poles. That ground wire up there helps provide lightning protection for the power lines.

For a lightning bolt, the "other pole of the battery" is the Earth because a lightning bolt is the result of a difference in charge between the clouds and the Earth.

Dirt is not infinitely conductive. In fact, in most places, it is a rather poor conductor. Another way to say that is, dirt is a resistor. Resistors create voltage differences between one side of the resistor and the other. *It is the difference in voltage between one place and another that creates the current that does the damage.* A typical lightning bolt is packing around 30,000 to 50,000 amperes of current at around 30 million volts. That's a lot of energy to dissipate, and it must be dissipated in a very short time – microseconds.

Regardless of the type of grounding, the purpose of a ground system is to provide *a minimum impedance path to ground* for whatever energy is being grounded.

The reason we want a minimum impedance path to ground is not exactly what most people think. Many hams I talk with about this have a mental picture of the Earth as some sort of massive electrical energy sink, as though once electrical

Figure 6.1: Equivalent Circuit of Ground

energy can find its way to some dirt, it sort of vanishes. It doesn't. The Earth is just another conductor. A big one, admittedly, but it's a conductor. More to the point, it's not a very good conductor; it's a big resistor, and resistors create voltage differences. Voltage differences are what create current, and current is what does the dirty work in all three grounding types.

There is a chilling scenario that illustrates this. Normally, when a power pole is damaged and a high-voltage power line falls to the ground, the line is quickly de-energized by a circuit breaker up the line at the nearest power substation. In some situations, though, that circuit breaker does not de-energize the line, and the line must be switched off manually. That means for some time, that power line is resting on the ground, charged with high voltage. That voltage does not magically disappear into the ground, it's radiating out through the ground around where the power line is down. There have been instances of people being killed because they got out of their car at the scene, took a step or two, then were shocked to death because the voltage on the ground at their right foot was significantly different – to the tune of 1,000 volts or more – from the voltage at their left foot.

We could represent an analogy of the above scenario with what the textbooks call an "equivalent circuit," like the one in Figure 6.1.

The real voltage of most neighborhood overhead power lines is actually closer to 13,800 volts, but we'll call it 10,000 to make the math easy. The resistors represent distances. Again, just to make the math easy, we'll say each resistor represents 10 feet from the downed power line. Kirchoff's Law tells us each resistor has 2,500 volts across it. A two-foot difference along the ground, then, has 500 volts across it.

We could eliminate all of these voltage differences by simply connecting jumper wires across each resistor, like Figure 6.2.

If it was possible, we could render the area around that downed power line safe by somehow creating "jumpers" every 10 feet or so to eliminate the voltage differences between one place and another. Those jumpers would create a minimal impedance path to ground. Of course, that's not practical once the ground is

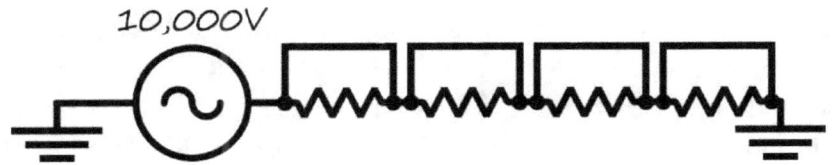

Figure 6.2: Equivalent Circuit of Effective Ground System

energized.

If we, for some strange reason, wanted to create the schematic in Figure 6.2 on a real printed circuit board, we'd probably just attach some wire jumpers across each resistor with some alligator clips. (See Figure 6.3.) When we're talking about the actual Earth as a series of resistors, we make our electrical connections by driving conductive stakes into the ground and electrically connecting them together with thick wire. In other words, we install ground rods and bond them together.

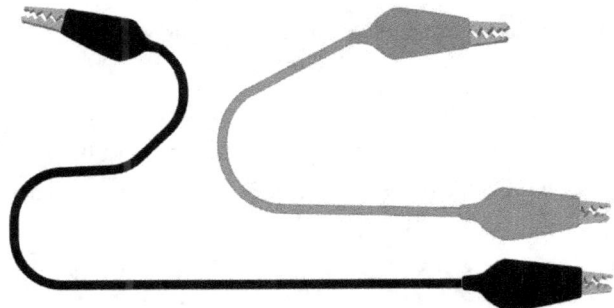

Figure 6.3: Jumpers with Alligator Clips

Meanwhile, back at the ham shack, if we can eliminate all differences in ground potential between our shack's ground and the house ground, we won't suffer ills that range from ground loops that create hum and noise to lightning surges that can burn out our house wiring. (If there's no difference in electric potential between your ham shack and the house ground rod, that lightning surge won't flow through that skinny #14 awg green ground wire in your outlet to the house ground rod.)

This objective of eliminating, or at least greatly reducing, voltage differences is why the license exams emphasize "bonding" all grounds together, which means to electrically connect them. It's that bonding that helps eliminate those voltage differences and thus eliminate harmful currents. The bonding is the wire "between the alligator clips."

By the way, electricians say "bonded" instead of "wired together" because bonding isn't always done with wire. When a metal electrical conduit system is installed, for instance, the bonding material is mostly the conduit itself. Any electrical discontinuity in the conduit system must be bridged by a piece of wire known as a bonding jumper.

Ground Rods

The base of your entire station ground system consists of one or more "ground rods." Ground rods are available at most home improvement stores. They are eight-foot long, copper-clad steel rods, usually one-half or five-eighths of an inch in diameter.

The rods are fairly inexpensive, and it is possible to install them yourself, though I won't claim it is easy.

The purpose of a ground rod is to conduct electricity into the soil. In the event of a lightning strike, that will mean conducting a lot of electricity into the soil in a very short time.

Each ground rod has a limited capacity. If it is a considerable distance from your shack to the house's main ground rod, you may need multiple ground rods to make the run; check your local code, but a common recommendation is that they be placed not less than six feet apart and not more than double their length apart. That means they will be sixteen feet apart unless you decide to drive ten-foot ground rods.

If your ham shack and antenna are located near your home's ground rod, then that ground rod represents the lowest impedance path to ground, and you can bond all grounds to that ground rod, so long as everything is within sixteen feet of that ground rod. Chances are slim that that's the case, though, so you'll probably need to add at least one ground rod to your system.

The rods must be driven into the ground. Let me emphasize; *driven*. There are some YouTube videos of folks using other methods, including using a garden hose to make a hole in the ground. That is *not* a National Electrical Code approved method, and for good reason. There's a very good chance that the wet ground left behind is going to shrink away from the ground rod as it dries; that won't be a solid electrical connection to ground.

If you are extraordinarily good at swinging a sledgehammer at full force while standing on an eight-foot ladder, I guess you could install your ground rod(s) by hand, but it sounds like a recipe for disaster. Those rods feel stiff right up until a

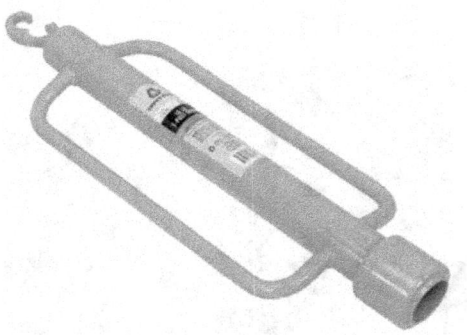

Figure 6.4: Fence Post Driver

hammer hits them, then they start whipping around all over the place. A better plan is to use a fence post driver like the one in Figure 6.4, but that's still challenging; probably fine if you're in good shape and driving just one rod, less fine for driving more than one.

To drive our ground rods, we used a Bosch power hammer with a ground rod driving "bit" in place. (Figure 6.5)The whole unit rents for around $125 - $150 per day in our area. Because our ham shack is on the other side of the house from our power service panel, we needed to drive six ground rods to go around the house, plus another three for our HF vertical antenna project, and had it all done in a half day. If I were doing the project again, I'd get a friend to help me manage that power hammer – the thing is very heavy and managing it while balancing on the ladder was much less safe than it could have been with some help.

All those ground rods are bonded with #6 awg bare copper stranded wire, all the way back to the original ground rod for the house. Figure 6.6 shows a schematic diagram of the system. The dotted line represents the bonding wire and the ground rods are shown by *'s.

The house ground rod at our place is located under the electrical panel. There's a ground wire that exits the ham shack and connects to that first ground rod. The lightning grounds for our dual-band and HF antennas also connect into that series of ground rods.

Safety Grounding

If something shorts out in your equipment, or some problem in your setup is causing high voltage RF or line voltage to be present on your equipment, you very

Figure 6.5: Driving Ground Rod with Power Hammer

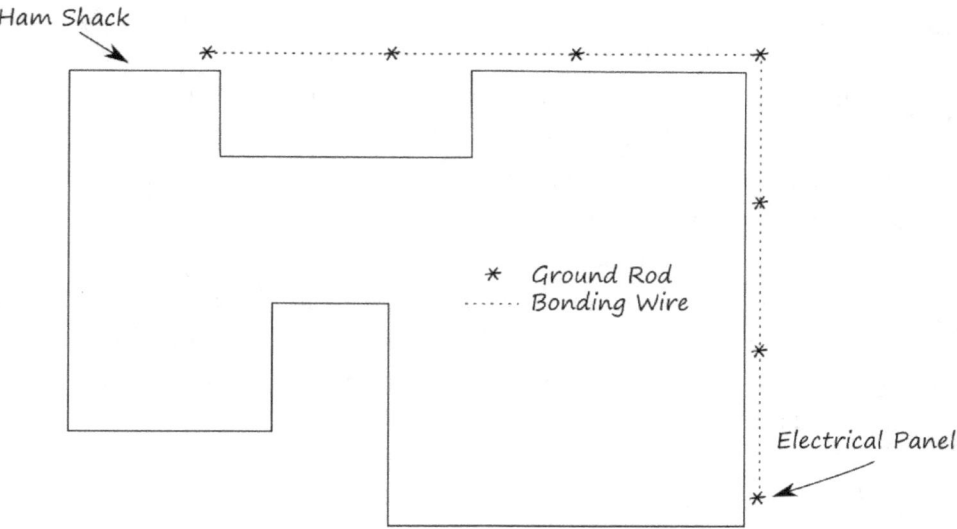

Figure 6.6: Schematic of Ground System

Figure 6.7: One-inch Grounding Braid

much want that voltage to go harmlessly to ground (tripping a circuit breaker or GFCI in the process) and not harmfully to ground through *you*.

Your power supply should be equipped with a three-prong power plug. The round prong is the one that connects to your home's ground wire. You might notice it is just a bit longer than the other prongs, so that it makes contact before power is applied to whatever is at the other end of the cord. That takes care of safety grounding for the power supply; or so you hope. However, some common ham radio power supplies do *not* connect the outer case of the supply to that ground. In that case, you should find a small screw marked "ground" (or with a schematic symbol for "ground") on the back of the power supply. That should be connected to your station's ground system.

What about for the radio? There's no three-prong plug on that. There may be a grounding screw on the back, though, and if so, that can be connected

While your transceiver operates on 12-volt DC power, it does contain high voltages in the final amplifier. It needs to be safety grounded.

RF Grounding

In addition to minimizing all the bad stuff that stray RF causes, good RF grounding has a side benefit; it helps keep environmental noise out of your equipment.

Once your grounding infrastructure is in place, RF grounding is mostly a matter of connecting all your grounds to the ground system with the correct conductor.

In the case of RF grounding, the correct conductor is very important. The ideal conductor is what's known as "one-inch grounding braid," which you can see in Figure 6.7.

Grounding braid is made specifically for RF grounding. The only manufacturer I've managed to find is ABR Industries, and the ham radio dealer for ABR Industries is Gigaparts. Grounding braid is not inexpensive; a 15-foot piece costs around $28 as of today. The good news is you only need enough to connect your shack's "grounding bus" to the nearest ground rod, plus enough short pieces to connect all the radio gear in the shack to the grounding bus.

A grounding bus is a hunk of copper plate with, usually, wing nuts mounted through it to secure the eyelet connectors you see on the grounding braid above.

If it isn't more than, say, twenty feet from your shack to the nearest ground rod, you'd probably be well grounded if you used #4 awg or even #6 awg stranded copper grounding wire for your RF ground. In that case, you can use a ground bus bar like the one in Figure 6.8 for your ground bus.

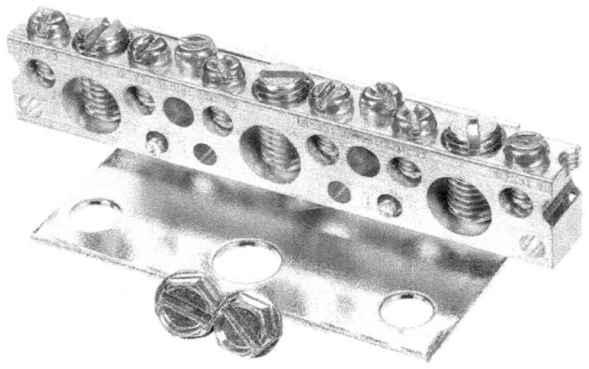

Figure 6.8: Ground Bus Bar

That ground bus bar is a component you can probably find in your local hardware store's electrical department. You can see it accommodates various sizes of ground wires.

Why all this emphasis on the correct conductor? Remember, the purpose of a ground system is to provide a minimum impedance path to ground. Impedance is resistance to AC current flow but it isn't quite the same as the resistance you know from Ohm's Law. Impedance can change depending on frequency, while resistance is always the same. You might measure the *resistance* to ground with an ohmmeter and think, "Gee, almost zero ohms, no problem there," but an ohmmeter uses DC to measure resistance. If, for some reason, your ground wire has high impedance to RF it will resist the flow of that RF, meaning that stray RF isn't getting to ground where it's supposed to be and is instead lurking on your equipment waiting to burn your finger, which is what happens if your RF ground is faulty.

How could you possibly end up with high impedance on a simple wire that goes directly to ground? One way to end up with undesirable high impedance would be to use small wire. RF likes relatively big conductors; small wires can present a high impedance to the signal. That's why it's recommended to use nice, beefy grounding strap (solid copper) or braid for your RF grounds, or, at the very least, #6 awg copper wire.

You may have noticed I've specified *copper* wire more than once. The alternative would be aluminum wire. In the short term, aluminum wire has higher resistance than copper wire. In the long term, aluminum wire is more prone to corrosion than copper, especially when it is connected to other metals. Choosing aluminum wire for your grounding would probably qualify as "better than nothing", but far from optimum.

Another important part of keeping RF out of your shack which is never mentioned in the exams is the connection to your antenna. Some antenna designs require what's called a *balun* at the feed-point, and may also require a "choke", which consists of several turns of your feedline wrapped into a coil. ("Balun" is a specific term that comes from "balanced to unbalanced" but in ham usage has come to refer to a wide variety of matching devices.)

Fighting Common Mode Currents

As you're setting up your antenna and grounding systems, you'll want to keep in mind preventing RF interference with your own signal. Part of that is accomplished with good RF grounding, part by judicious placement of chunks of ferrite known as ferrite chokes.

RF interference typically rides into your transceiver as what is known as common mode current.

To understand common-mode current, it's useful to understand its opposite, differential-mode current.

Differential-mode current is the current we normally expect in an audio cable or an antenna feed line, or even the cord to your table lamp; equal and opposite currents flowing in the two sides. To make it easy to picture, let's imagine a couple of instants in the life of a piece of ladder line, like the one in Figures 6.9 and 6.10.

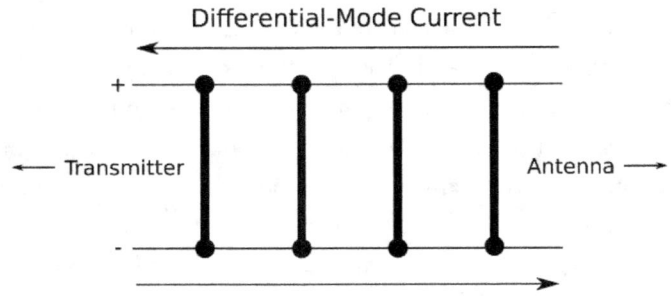

Figure 6.9: Differential-Mode Current

In Figure 6.9, the top wire is positive, the bottom wire is negative, and current is flowing at equal rates in each, but in opposite directions. All is well! There's no "excess" energy in the feed line. Not only that, that feed line will not radiate any signal because the electromagnetic waves from one side are canceled by the waves from the opposite side. This is good; all our signal is getting to the antenna (or to the receiver – this works in receive mode, too.)

Technically speaking, common-mode current is any current that is not equal and opposite on both sides. Figure 6.10 shows another instant in the life of that ladder-line, and in this instant, there is a lot of common-mode current – current is flowing in the same direction in both wires.

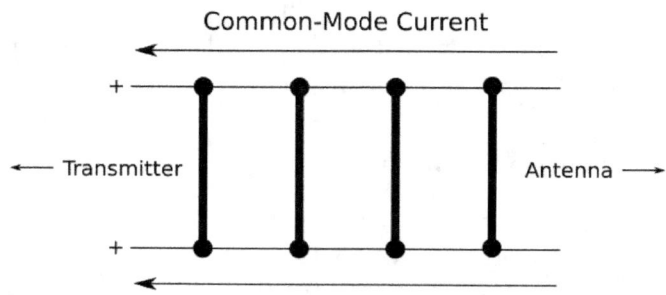

Figure 6.10: Common-Mode Current

Now that common-mode current is looking for a place to go. If the common-mode charges on each side of the feed line – or audio cable – aren't equal, at least some current can flow from one to the other, and then we experience interference.

Ideally, the common-mode current finds a safe path to ground and is out of the system or hits some sort of impedance and gets diminished to the point that it no longer matters. If it doesn't – it will find its way to your finger and give you an RF burn when you reach over to adjust the power level on your transceiver while you're transmitting.

How can this common-mode current stuff happen? Quite a few ways! First, if your antenna is within one-half wavelength of your equipment, it can drive a lot of RF into your gear. Engineers say it can "couple" RF into the system. In the ladder line's case, equal voltages were induced in both wires by the strong RF field from our station. Since not many of us have the land to put an 80-meter antenna a full 130 feet away from our shack, we must take other measures to suppress the common-mode current. That's why when you go to a hamfest you'll find vendors selling bags full of ferrite chokes.

That piece of ferrite won?t have much effect at all on the differential currents

Figure 6.11: Snap-on Ferrite Choke

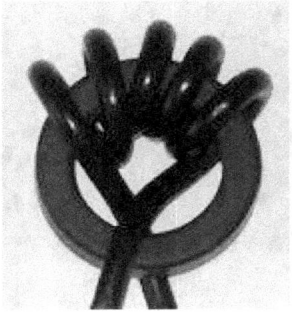

Figure 6.12: Ferrite Ring

in the center conductor and shield of a coaxial cable. Their electromagnetic fields are canceling each other, so they won?t interact with the ferrite to create inductance. Their electromagnetic fields are canceling each other, so they won't interact with the ferrite to create inductance.

Since the common-mode current is all headed in the same direction, it will generate an electromagnetic field, and that field's interaction with the ferrite bead will create inductance and, thus, impedance for the common-mode current. (Inductance is a component of impedance.)

Anything with inductance can be referred to as a choke – it "chokes" the current, especially high frequencies like RF. The easiest ferrite choke to apply is the snap-on type. You just open it up, then snap it shut around the wire. Figure 6.11 shows what one looks like.

If that doesn't do the trick, you might need more wire around the ferrite, in which case you might get yourself a ferrite ring (or toroid) and create something like the ferrite ring shown in Figure 6.12.

Ferrite has high *permeability*; formally, that's the ability of a substance to support the formation of a magnetic field. For us, it's a measure of how that substance will affect an inductor's inductance. Simply put, if you put a chunk of ferrite next

Figure 6.13: You Don't Want This in Your House

to an inductor, that inductor's inductance will become much higher because of that permeability.

Lightning Protection Grounds

What is, perhaps, the most important feature of your station is one you might hope you will never use; your lightning protection system. Happily, a good lightning protection system will be working for you even on those days that the sky doesn't single you out for a special delivery thunderbolt.

There is no state in the United States where your risk of a lightning strike is zero. There are no cities with zero thunderstorms per year. If you have a radio antenna, you're at risk. If you happen to be in the Pacific Northwest or Alaska, you're at less risk than if you're in Texas or Oklahoma, but you're still at risk.

It's prudent to take some (fairly simple) steps to at least minimize the damage from a lightning strike. The truth is, it's almost impossible to provide 100% protection from *every* direct lightning strike. However, a bit of "wisdom" that goes around ham radio circles is that it's impossible to have zero damage from a direct strike. Nonsense. Power lines and commercial radio towers get direct hits every day of the week, and seldom take any damage at all because they employ well-engineered lightning protection.

What is far more likely to occur than a direct hit is a *nearby* lightning strike. Consider; an antenna's whole job is to capture faint electromagnetic waves. A nearby lightning strike's electromagnetic waves are not faint, they're very strong, so a powerful voltage will be induced in your antenna and your feed line will do an excellent job of conducting that voltage directly into the front end of your receiver. Not good.

Lightning Basics

Our big concern in the domain of lightning is cloud-to-ground lightning. Cloud-to-cloud lightning and intracloud lightning (lightning inside a cloud) can also affect our station, but the catastrophic damage we dread most is produced by cloud-to-ground strikes. Folks who study lightning talk about "CG," "IC," and "CC" strikes.

A CG lightning bolt is the result of a difference in charge between clouds and the Earth. When the difference in charge becomes great enough, it overcomes the insulating nature of the air and an arc is the result. A single lightning strike begins with a voltage between the cloud and Earth of more than 100,000,000 volts. The peak amperage of an average strike has been estimated at around 30,000 amperes. If we had to dissipate a sustained current of 30,000 amperes, we'd almost certainly fail, but the lightning bolt only lasts about 0.2 seconds, and is actually made up of several pulses lasting only microseconds. In total, an average lightning bolt will conduct 15 coulombs of electricity – about what an ordinary household 15-amp outlet can deliver over the course of one second. Put that way, it doesn't seem like much, but it all arrives in milliseconds.

First, to review, *it isn't the voltage that does the damage, it is the difference in voltage between one place and another and the current that flows between those places.* Keep that in mind.

The Earth is not infinitely conductive. Far from it. The most conductive soil in the continental US is found in the middle of the country, in patches that cover areas in a belt that stretches roughly from North Dakota down into Oklahoma and North Texas. Even that soil only has a conductance of 30 millisiemens per meter, so a resistance of 33 microohms per meter. By way of comparison, seawater has a conductance of 3,000 millisiemens per meter – 100 times the most conductive soils and amounting to about 333 nanoohms per meter; practically no resistance at all. Most of the country is far lower than 30 millisiemens – here in my neighborhood in Western Washington, the ground conductance is about 2 millisiemens, so a resistance of a one-half ohm per meter. You'd think it would be far more conductive, since we get so much rain and the ground stays soggy much of the year, but water is a terrible conductor unless it contains minerals, and neither our rainwater nor our soil contain much in the way of dissolved minerals.

What that means is that any given patch of dirt can only dissipate a certain amount of a lightning bolt. Beyond that, the resistance is too high and the lightning will start looking for an easier path to ground.

Ground rods are not infinitely conductive, either. There's a limit to the quantity of electricity a ground rod can conduct into the Earth. That's why almost every well-designed ground system incorporates multiple ground rods. Rods spaced too closely together interfere with each other – remember, the ground can only carry so much current. A system with rods spaced too far apart won't conduct as much current as it could. According to *Electrical Construction & Maintenance* magazine, the optimum spacing is approximately double the length of the rods; for 8-foot ground rods, 16 feet.

By the way, 8-foot rods are *not* standard everywhere for every application. Check your local code. In Florida, for instance, with notoriously sandy, non-conductive soil, lightning protection specialists routinely drive 20-foot ground rods, which consist of two 10-foot ground rods joined in the middle.

Of course, all the ground rods need to be electrically bonded – which means they're all wired together with a big ol' # 6 or #4 copper wire – so what doesn't get dissipated by ground rod #1 gets dissipated by ground rod #2, etc., and so the ground rods and the ground around them all stay at the same potential.

Think of the ground rods and the ground surrounding them as parallel resistors – the more resistors you have connected, the lower the resistance and the greater the current flow.

Because I've encountered so many folks who were confused about this, I'll emphasize again, *all* your ground rods get bonded. There's no separate lightning protection ground system – it's all one ground system.

Figure 6.15 is an illustration of the home of a ham who got the station's grounding system *almost* right. You can see there's a ground rod at the service entrance, where power lines enter the house. There's also a ground rod outside the ham shack, and one for the (modest) tower.

What's missing? Bonding! Each ground rod is isolated from the other ground rods.

Let's think this through. If lightning hits the tower, some of it goes to ground through the tower's ground rod. Over there in the ham shack, the shield of the

Figure 6.14: US Airman Installing Lightning Protection - Note size of bonding wire.

feed line is connected to the ham shack's ground rod. Now that ground rod and its surrounding dirt are at a lower potential than the ground rod at the base of the tower, so some of that lightning bolt's energy goes through the feed line to ground and the rest – uh, oh – goes through the radio to the house wiring, then through the house wiring to the ground rod at the service panel because the ground at the service panel is at lower potential. We just sent a few thousand amps through wires that were designed for 15 or 20 amps.

If, instead of the tower, lightning hits the power lines – a far more likely scenario, since there are miles and miles of power lines and only your single tower – the same current flow is going to happen in reverse. (By the way, that tells you how futile the often-repeated advice, "just disconnect your feed line during a thunderstorm" is – no feed line is necessary for that unwanted current flow to happen.)

Now consider the situation with the ground rods bonded, as in Figure 6.16.

When lightning strikes the tower or the power lines, the bonding between the ground rods almost instantly puts all the surrounding ground at the same voltage. Current flow is minimal and, with a little luck, everything's safe. At worst, damage is minimized.

"But if my radio is connected to the same ground system as the tower, won't the lightning try to come into my radio?" It's coming anyway! It will come through

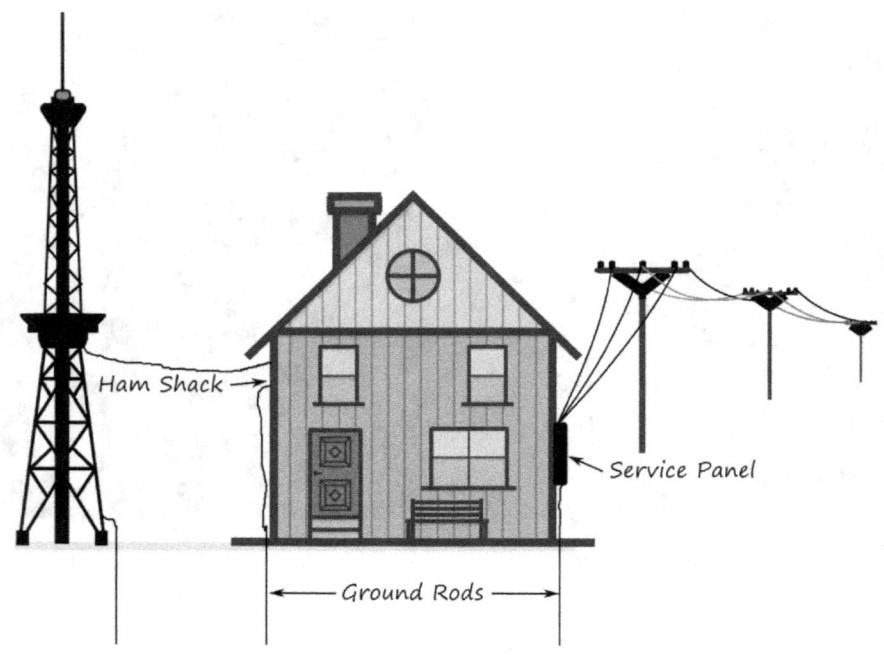

Figure 6.15: An "Almost Right" Ground System - Don't Do This!

the ground system or it will come through the Earth, but it's coming your way. If it encounters a properly designed and installed ground system there's no voltage difference between "ground" and your radio's ground, because they're the same thing. No voltage difference, no current flow. (That said, be prudent and disconnect your coax and power cords *before* a lightning storm – once you hear thunder, though, it's too late and definitely not safe to be handling your coax.)

The shield of your coaxial feed line needs to be grounded, too. If you have a PL-259 in a convenient place you can clamp a ground wire onto it with a stainless steel hose clamp around the connector, or just strip a couple of inches of the insulation close to the nearest ground rod, apply a clamp, and wire the shield to the ground rod. You can – and should -- also install a *lightning and surge protector* in the feedline; that device also needs to be connected to ground.

Do you really want lightning in your house? (No. You do not.) Then DO NOT bond your lightning protection grounds to water or, even worse, gas pipes! Those pipes are great conductors of electricity and they're going into your house, the neighbor's houses, and who knows where else. Besides, that copper or iron water line in your home is probably plastic after it leaves your house –contractors aren't going to pay for copper pipe all the way to the water main when plastic works

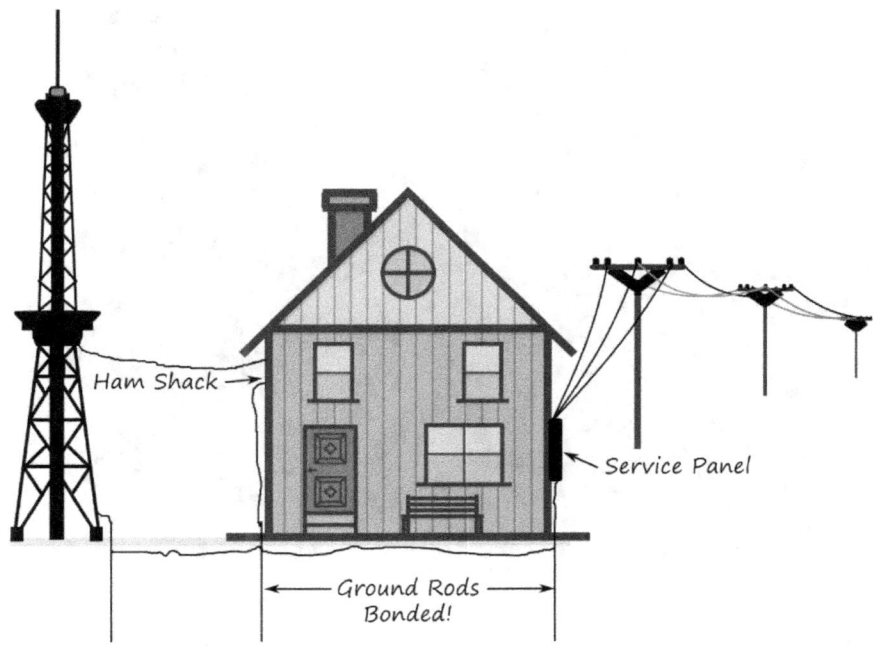

Figure 6.16: Proper Ground System - Do This!

just as well, if not better. The gas pipes are full of stuff that goes boom when it gets some oxygen and a spark, and it's not too hard to imagine the energy of a lightning strike melting through some 3/4" iron pipe. Let's see -- leaking gas pipe plus lightning spark -- what could possibly go wrong? Oh. Everything.

DO NOT bend ground wires at sharp right angles. Bends in lightning protection ground wires should be avoided completely if possible, but if they are unavoidable – and they usually are -- the bends should be gradual curves, certainly not sharp right angles.

DO NOT use solder on any grounding connections. Use only approved bonding clamps, usually sold as "ground rod clamps." In the heat of a lightning strike, solder will melt instantly and the void it leaves behind will create a poor electrical connection at the precise moment you need a perfect electrical connection.

Figure 6.17 shows the top of one of our ground rods.

The top of the ground rod is about an inch below the soil level, to avoid creating a trip-and-fall hazard. The bonding wire is buried a couple of inches below the top of the soil. That clamp is a bonding clamp, and it's available at most big-box home improvement stores for a couple of bucks.

If you need to join two bonding wires, you use a split bolt, like the one in

Figure 6.17: Ground Rod with Bonding Clamp

Figure 6.18.

It doesn't take a direct strike to damage your equipment and your home, either. You're also protecting your house and your gear from nearby strikes; the voltage induced in your antenna and transmission line by a nearby strike can send a lot of volts to bad places. In fact, just the static charge caused by wind blowing past your antenna is enough to damage the front end of your transceiver. This is what I meant when I said the system would be working for you every day.

Ground Loops

If you have hum on your signal, chances are good you have a ground loop. Ground loops are the result of failing to electrically tie all station grounds to the same physical point, which is typically a ground rod. A ground rod is a steel or copper-clad steel rod, usually eight feet long, driven into the ground.

What causes a ground loop? My friend who spent years of his career as a grounding specialist at a telecom company told me, "It's simple. Not all ground is ground!" In other words, not all earth is equally conductive, so different ground points are almost certainly at different potentials. Because they are at different potentials, current is flowing where current should not be flowing. That current gets modulated by stray RF or AC hum. Now you have inadvertently constructed a big loop antenna that does a great job of picking up all sorts of noise from the environment – particularly that 60 Hz hum, which is everywhere.

Figure 6.18: Split Bolt

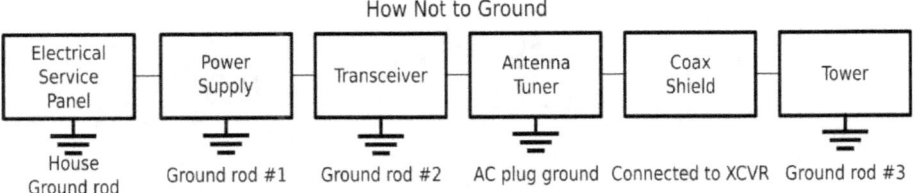

Figure 6.19: How to Create a Ground Loop

Figure 6.19 shows how NOT to set up the ground system for your station.

In Figure 6.19, our paths to Earth are five separate ground rods, presumably widely separated from each other and not bonded. Chances are very good that if you heard that ham's signal, it would have a big, nasty 60 Hz hum on the audio, because that set-up is a perfect recipe for ground loops.

If all those grounds were tied together with a low impedance copper wire – the bigger the better – they'd all be at least closer to the electrical equivalent of a single point. It's bonding again! (See Figure 6.20.)

It would be even better (at least in regard to ground loops) if we could get everything tied to a single ground rod, but as we covered back on page 106, there are also limits to what any ground rod can conduct and limits to the ground's conductivity. Again: Check your local code for how closely ground rods should be spaced in your area.

Any time you have hum -- even in non-ham equipment -- suspect a ground

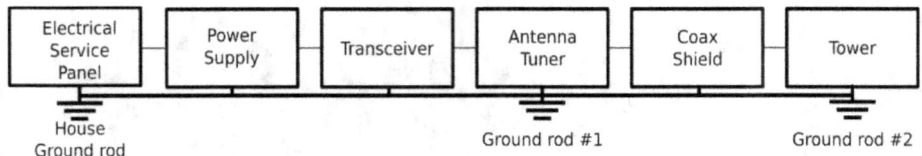

Figure 6.20: No Ground Loop

loop somewhere. Skilled car stereo installers know this very well, and a high-quality car stereo installation will bring all the audio gear grounds to a single point, ideally right at the battery or, at least, the engine block. You should follow the same practice when you install your mobile rig in your vehicle and follow the same principle when you set up your home shack.

Chapter 7

Operating on VHF/UHF

Dual-band Radio Controls

Compared to HF transceivers, most dual-band radios have relatively few controls. Chances are good you'll only use a handful of the available controls on a day-to-day basis.

- **Volume**. This one's pretty self-explanatory but be aware that the volume control only affects the received audio – it doesn't adjust your transmit volume. If there's a control for that – and there probably isn't – it's called "mic gain."

- **VFO**. In plain English, the VFO (Variable Frequency Oscillator) is the tuning knob. On most dual-band radios, it doubles as a memory slot selector, depending on if you have selected the VFO function or a function named something like "MEM."

- **Squelch**. This may be a new one for you. Squelch keeps the radio muted unless it "hears" a signal. That way you aren't constantly serenaded by static. On many handheld transceivers, squelch is set through the menu program. On most mobile/base transceivers, it's often a ring around the volume knob. The way to set it for normal use is to lower it until you hear static, then raise it just enough to shut off the static. If you're trying to hear a really faint signal, you might switch off squelch altogether.

- **Reverse**. One slightly advanced trick for repeater operations you'll want to know about is "listening on reverse," also known as "listening on the input."

Some radios have a handy way to listen to the input frequency of the repeater. Activating the "reverse" function flips your "listen" and "transmit" frequencies. This comes into play if someone's signal doesn't get into the repeater strongly. It's worthwhile to activate your reverse function while they're talking; you might hear them better if they are near you. For instance, there are a few other hams in our neighborhood who regularly participate in our club net and whose signals come in much better at our house on reverse than it does from the repeater. That might not be true for the person directing the net, the "net control station", so sometimes net control will say something like, "I'm not quite copying the station that was just transmitting – did someone get them on the reverse?" If you did, say your call sign (remember to deactivate reverse!) and wait for net control to acknowledge you, then relay what the unheard person had to say as well as you can.

This is certainly not a complete list of every control that can be on a dual-band radio; my Kenwood mobile has all sorts of bells and whistles, but those are "once in a blue moon" controls for the most part.

Repeaters

Let's say you have never made a contact using a ham radio. How does one get started?

Working through a repeater is probably going to be the easiest way to make your first contacts. Good repeaters give you greatly expanded range compared to a handheld or mobile radio. The benefits of that expanded coverage include a greater likelihood that someone else is within range and listening as well as, most likely, improved reception on your end for easier intelligibility.

A repeater is created by combining a receiver with a transmitter. In the simplest form of repeater, the transmitter simply transmits everything the receiver "hears." The transmitter must transmit on a slightly different frequency than the one the receiver is listening to – otherwise, the transmitter would just wipe out the incoming signal (and probably burn up the first stage of the receiver.)

Receiver sites are chosen to give good coverage. Preferred sites for the antennas are the tops of mountains, the tops of buildings, or even sharing space on a commercial tower. Because of these advantageous locations, repeaters turn your little station into a big station.

		NATO Phonetic Alphabet					
A	Alpha	J	Juliet	S	Sierra	1	One
B	Bravo	K	Kilo	T	Tango	2	Two
C	Charlie	L	Lima (LEE-muh)	U	Uniform	3	Three
D	Delta	M	Mike	V	Victor	4	Four
E	Echo	N	November	W	Whiskey	5	Five
F	Foxtrot	O	Oscar	X	X-Ray	6	Six
G	Golf	P	Papa	Y	Yankee	7	Seven
H	Hotel	Q	Quebec (key-BECK)	Z	Zulu	8	Eight
I	India	R	Romeo	0	Zero	9	Niner

Table 7.1: NATO Phonetic Alphabet

When you transmit through a repeater, you need to give the repeater time to open its squelch when it hears your tone, then for the transmitter to key up. That usually takes about a second, so after you key the microphone, count "a thousand and one" to yourself before speaking. That way, your first words – which will usually include your call sign – won't be clipped off. That little bit of dead air hiss at the start of your transmission also alerts people on the other end that someone is about to speak. Think of it as the radio equivalent of tapping a glass with a dinner knife to get everyone's attention for a toast.

Use the phonetic alphabet as shown in Table 7.1. You'll need to be comfortable with both sending it and receiving it, and that takes some practice. One exercise is to practice translating things like stop signs or web site addresses into phonetic. A stop sign becomes a chance to practice "Sierra Tango Oscar Papa." Watching a YouTube video lets you practice "Yankee Oscar Uniform Tango Uniform Bravo Echo." This relates to what can be a bit of a bugaboo for beginners; hearing and repeating other hams' call signs. If you find yourself struggling with that, just do the best you can. It's fine to just use what you caught – I have been called "7KB" plenty of times. When I was KG7DVV it was a very rare occasion that *anyone* got my call sign right!

A useful accessory when you are starting is a notebook and pen. Use them to jot down call signs and names as you receive them; it really helps and even experienced hams use them a lot.

One time-honored method to make that first contact (or any contact) is what's called, in ham speak, a "sked." It's a scheduled contact. It requires knowing another ham. You arrange something along the lines of, "At ten o'clock this morning, I'll call you on the 146.920 repeater." At ten o'clock, you push the PTT button, say their call sign, then your call sign. They'll answer back with your call sign, then their call sign. They'll let you know how your signal is being received, and you'll do the same for them. Any other conversation is up to the two of you.

You don't have to wait for a sked, though.

You could just pick up the microphone and say your call sign and the word "monitoring," then listen for a response. If I was doing it, it would go like this: "AF7KB, Alpha Foxtrot Seven Kilo Bravo, monitoring." If I was mobile, I could say "mobile" instead of monitoring, just to let people know I was en route to somewhere. I could even say "mobile and monitoring." I could just say, "listening." Whichever way you choose, the message you are sending is, "I'm here at the radio and open to any conversation."

That's a little passive, though. You could also listen until you hear another conversation on the repeater end, then pick one of the hams, say their call sign, then your call sign, which is short for "Hey AF7KB, this is KC7YL," and start a conversation. That, however, depends on you hearing that other conversation – on some repeaters, you could wait a long time for that! Besides, maybe that feels like you're being a little forward. You're not – hams who turn on a radio are looking for a conversation; but I understand the feeling.

Instead, try asking for a signal report, whether you're tagging on to the end of a conversation or there has been nothing on the repeater for hours. You do want to know how well your radio is transmitting, right? Once you have your radio programmed, you simply key the mic and say, "This is [call sign.] Can someone give me a signal report, please?"

I heard a signal report exchange yesterday that is very typical. I'll give the hams fictitious names and call signs; "James" is WW7ABC and "Ted" is KK7XYZ. Ted had been listening to a conversation between James and another ham on our club repeater. When that conversation ended, things went like this:

Ted: WW7ABC, this is KK7XYZ. My name's Ted and I'm in Everett (Washington) – how's my signal doing today?

James: KK7XYZ, this is WW7ABC. Hi, Ted, the name's James. You're loud and clear here in Arlington (Washington) – yeah, very clear, full quieting (that means the transmission was free of static) and nice audio level. No problem copying you (understanding what you're saying) at all.

Ted: Oh, great. I'm on my new handheld and I'm on full power right now, so five watts. I'd like to switch to two watts and get a report if you have a moment.

James: Sure, go ahead!

Ted: Okay ... It will take me a second to switch power. Standby.

James: Standing by!

Ted: WW7ABC, this is KK7XYZ on two watts. Do you read me?

James: KK7XYZ, I can sure tell you switched power – there's quite a bit of static now. I can still copy, but, yeah, lots of static. You sounded like you were right next door before.

Ted: (Still on two watts of power) Darn! I was hoping this would work okay on two watts – that sure saves the battery life.

James: Yep, it sure does. No trouble understanding you, just some static on the signal, but copy just fine.

Ted: Okay. Thank you very much, James. I think when I'm operating from here, I'll go ahead and set it on five watts. I don't want to be known as the Static Guy!

James: (laughing) Yeah, that would be terrible! Like I say, though, I could copy you fine, even at two watts. Hey, no problem, any time you need a signal report feel free. Good talking with you!

Ted: Thanks, James. KK7XYZ clear.

James: You're welcome, Ted, WW7ABC clear.

(My apologies if either of those call signs gets assigned to you – at the time of writing, those were both unassigned.)

Notice that there was a bare minimum of unintelligible jargon in the conversation. That's typical of VHF/UHF communications in particular, but, really, it's good practice on any band.

Just before Ted called, WW7ABC had been in a short conversation with a ham who was commuting home on the I-5 freeway from north of here. That conversation was about where each of them was headed – they were both operating mobile – how traffic was where they were driving, and what repeaters were the most active in the area.

You can see that the topics of conversation in ham radio are often something to do with ham radio, or just normal conversation.

There's no guarantee that someone will answer your request for a signal report the first time you try. There may be no one listening, or those listening may not be able to respond at the moment, or just aren't in the mood for a chat. It may take several tries.

Nets

Another way to get into a conversation is to get on a local net. There are lots of them! You can find many nets through the ARRL Net Search.

http://www.arrl.org/arrl-net-directory-search

I just searched for local nets in Washington State on the 2-meter band and found 39 listed. However, I know for a fact that there are more than are shown, since none of our club's three weekly nets are listed. So, try that search, but also check with your local clubs.

Nets can meet on a daily basis, weekly, monthly, or even "as needed."

Some nets are almost completely informal and unstructured. Imagine a group that meets for coffee every Thursday morning at 11 AM and chats about whatever's on their minds. Some don't even have a particular time they meet; there's a local repeater that is often used by local hams who are truck drivers, and their "net" happens whenever two or more of them decide to sign on.

Other nets are tightly focused on a specific topic, with a very disciplined structure. In Seattle, for instance, there is a weekly net just for local leaders of the LDS Church, focused on church related topics.

There are "traffic-handling" nets. These are focused on practicing emergency message handling skills. Frankly, they often consist of a long series of, "This is [call sign]. No traffic," so they don't often make for very interesting listening for outsiders, but that's not what they're intended to be; those folks are practicing to be of service in times of emergency.

Then there's the sort of net you're looking for; a weekly club net focused on topics of general interest, or as at least one local net advertises, "well-being checks."

Nets are useful for the beginner if for no other reason than you know *someone* will be on the air at a particular time on a particular frequency. That person is known as "net control" and is the moderator for the net.

There are no FCC rules about how a net should be conducted. Every organization or individual who wants to run a net is free to do so however they choose (within the limits of Part 97, of course.) For that reason, I can't tell you precisely how the net you join will operate, but a well-run net will make clear to the participants when and how to participate. Here's how things might go: typical nets begin with a prologue that will be something like this.

Net Control: Good evening and welcome to the Anytown Amateur Radio Club's Sunday Evening Net for August 25. This is AF7KB, your net control for this evening's net.

This net meets every Sunday evening at 7PM here on the club's VHF repeater, WW4ANY, for the purpose of testing equipment, disseminating news, soliciting input, and for providing a forum for discussion of amateur radio related topics.

This is a directed net. Stations wishing to check in may do so by stating their call sign when net control asks for check-ins. Stations are asked to check in first and hold their traffic until instructed to go ahead by net control. Stations having emergency or priority traffic may break in at any time.

Are there any stations having emergency or priority traffic at this time?

In other words, don't talk yet, unless you have an emergency – then, talk any time you want. After a suitable pause, net control will continue.

If at any time during this net you wish to contact another station on the net, please say your call sign and wait to be recognized by net control.

If someone says something to which you wish to respond, you don't just key the microphone and respond directly. Instead, say your call sign and wait for net control to tell you to go ahead. This helps prevent "doubling" which is the unpleasant result of two operators transmitting at the same time.

Both club members and non-members are encouraged to check in and participate in this net.

Great! Not all nets have the same policy, so be sure you listen to the prologue before checking in to any net.

Let's start with a roll call of our net control stations.

Some groups rotate the net control duties among a group of volunteer net controllers and have a policy that those people get to check in first. You're not a net controller (yet) so let them go ahead.

At this time we will pause so all EchoLink stations have an opportunity to check in.

Some nets create a special check-in time for those who are calling in on EchoLink. That's the cell phone app that connects to EchoLink equipped repeaters. If you're using EchoLink, this is your moment to check in.

Now we'll start taking regular check-ins. Stations wishing to check in may do so now. Please state your call sign, name, and location. Let's start with stations having call sign suffixes beginning with Alpha through Echo. Alpha through Echo, go ahead.

It's common practice to open check-ins to a limited group of people, and the handiest way to do that in ham radio world is by "call sign suffix." The call sign suffix of my call sign, AF7KB, is "KB", so I would not check in at this time. Grouping the check-ins this way is another tactic to minimize doubling.

Once it seems that everyone with a call sign suffix beginning with Alpha through Echo has checked in, net control will continue through the alphabet. Often this goes *Alpha through Juliet*, then *Alpha through Oscar, Alpha through Tango*, and finally *Alpha through Zulu*.

With my call sign, I'll wait until the call for suffixes *"Alpha through Oscar."* Then I'll state what net control asked for. In this case, that's "Call sign, name, and location. *"AF7KB, Alpha Foxtrot Seven Kilo Bravo. Michael in Lake Stevens."* I might have a lot more to say, but this is not the time to say it. I'm just checking in.

As people check in, net control will acknowledge them, often with something like *"AF7KB, Michael I have you down."* Net control will be making a list of the call signs and any other related information they plan to use, such as the operator's name and location. If you don't get acknowledged, you probably won't be called on to participate in the net. Send your information again.

Net control is a more intense job than you might think, and they will appreciate you stating your call sign clearly and using the phonetic alphabet to make it doubly clear.

Almost inevitably during the check-in process, two or more people will transmit at the same time, creating a more or less completely garbled signal known as a double. Net control will straighten it out. Usually, there's some clue they can pick up about one of the senders, such as a name or location. In that case, they'll say something like, *"Okay, it sounded like we had a double there, so station in West Othertown, you try now, then whoever else was in there. Go ahead West Othertown."*

You might hear some people check-in and add *"In and out."* That means, "Just mark me as present. I don't have anything to say tonight, or I can't stick around for the whole net." Since people are often long-time regular participants in nets, this lets others know you are all right and they have no need to worry about your well-being.

Once everyone has checked in, net control will start calling on people in the order in which they checked in. When your call sign is called, that's your turn to share what you're going to share. The first time you check into a net, I'd suggest you introduce yourself as a new participant, thank net control for "taking the net", and thank everyone for letting you participate. Maybe share a little bit about how long you have been a ham or what you enjoy about the hobby. Remember to give your call sign at the end, the same as you would with any other conversation on ham radio, and then say, *"back to net."*

Once everyone on net control's list has had their turn, net control will probably ask if there are any *"missed or late check-ins"* for those who arrived after the first round of check-ins, and then wrap it up with their version of, *"Join us again next week, same time, same station."*

You can see that net control's job is to control the flow of communication and keep the net on track. You'll do well to listen to a full session of the net you have chosen before you jump in and participate.

KC7YL and I are on our club's Sunday night net just about every week. It has become a regular part of our social life. She's also on the rotating roster of net control operators for that net as well as net control for the Thursday evening YL (Young Lady – in other words, female hams) net on our club repeater.

The length of a net depends completely on the number of participants and what they have to say. There aren't any statistics, but my experience is that few last more than an hour. Some are brief – one night my wife did her opening announcement, waited for check-ins, and nobody showed up at all, so that one was *very* short.

EchoLink®

For times when you are out of radio range of any repeaters, there's still EchoLink®. EchoLink is a way to connect with EchoLink equipped repeaters through your cell phone or your computer. If you have a smart phone and a cell signal, or if you have a computer with access to the internet, you can get on the radio.

Not all repeaters are equipped with EchoLink. Enabling EchoLink on a repeater requires a computer – most likely one dedicated to the task. Internet access for the computer is also required. The software that runs EchoLink is free to download and operate.

It's likely that at this stage of your ham radio career, the usefulness of EchoLink is not instantly apparent to you. That will probably change as you become a regular member of various on-air communities and want to keep in touch. One especially memorable Thursday evening found us in Orlando, FL for the huge hamfest known as Hamcation; Kerry did her net control duties for the YL net via EchoLink from our hotel room.

You're probably most likely to use EchoLink from your cell phone. To do so, you download and install the EchoLink app from the Apple or Google Play store. There's an authentication procedure for you to complete to prove you are a licensed amateur. Over the years this procedure has gotten more streamlined, so I won't attempt to teach you how to complete it here since it will probably evolve some more, but even now it is a simple and certainly a necessary procedure; can you imagine the consequences if e-mail spammers could access ham repeaters?

Once the app is installed and you are verified, using EchoLink is very easy. There's a screen, Figure 7.1 that allows you to search for any EchoLink enabled

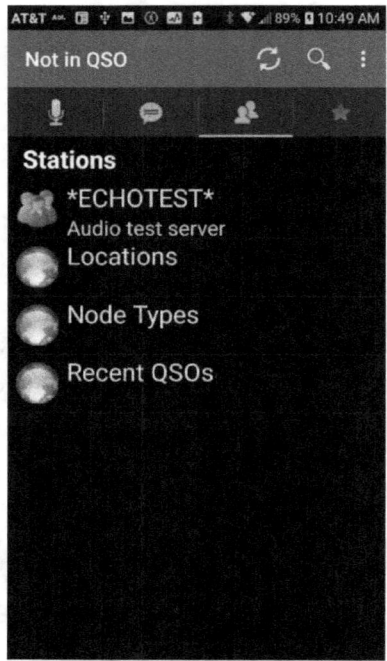

Figure 7.1: EchoLink® Home Screen

repeater.

You can see the search icon in the upper right section of the screen shown in Figure 7.1.

ECHOTEST is, as it says, an audio test server. If you've ever used Skype's audio test, this is very similar. You select that station by tapping it, then you can make a test transmission to the server. Nothing is going out on the air, so don't worry about ID'ing.

You also don't need to worry about the rest of the selections on that screen, at least for now – honestly, we've used EchoLink for years and have never used any of those selections.

Once you've located a repeater with which you want to connect, you can save it as a "favorite" by pressing the "Add to Favorites" button that will appear when you tap the repeater name in the search results.

When you have a repeater set as a favorite, you can go to the Favorites screen and tap the repeater name. When you see the screen that looks like the one above, just tap Connect and the app will log in to the repeater.

On the repeater end, a computerized voice will automatically announce your

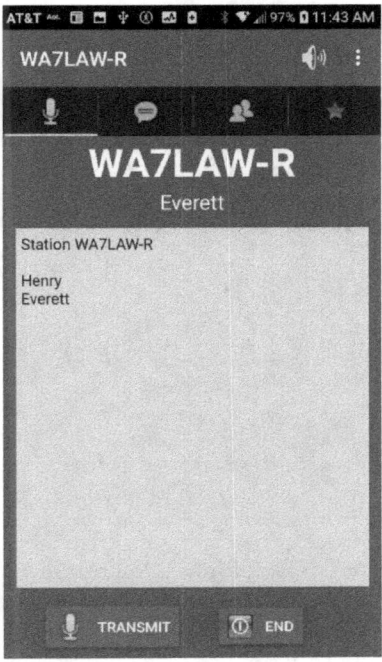

Figure 7.2: EchoLink® QSO Screen

call sign when you log on. When you disconnect, it will announce your call sign again, and say "disconnected." Your "end of communication" station ID is handled automatically, so in case you drop off unexpectedly, you're still legal. (You still need to handle the "every 10 minutes" part of your station ID requirements, though.)

When you connect with the repeater you'll see a screen that looks like Figure 7.2

In Figure 7.2 I've logged into our club repeater, WA7LAW.

You'll see a list of other EchoLink users who are presently logged into the repeater. In this case, you can see someone named Henry, who is in Everett, WA, is logged into our club repeater. (Henry's almost always logged on to EchoLink from his home computer, because he keeps an eye on all our repeater's computer functions.)

You'll hear it over your phone if anyone transmits on the repeater, whether they are using EchoLink or not.

At the bottom you can see a Transmit button – that's your push-to-talk button. Other than your phone's volume control, that's it for controls.

There's a program for your desktop computer, too, shown in Figure 7.3, but

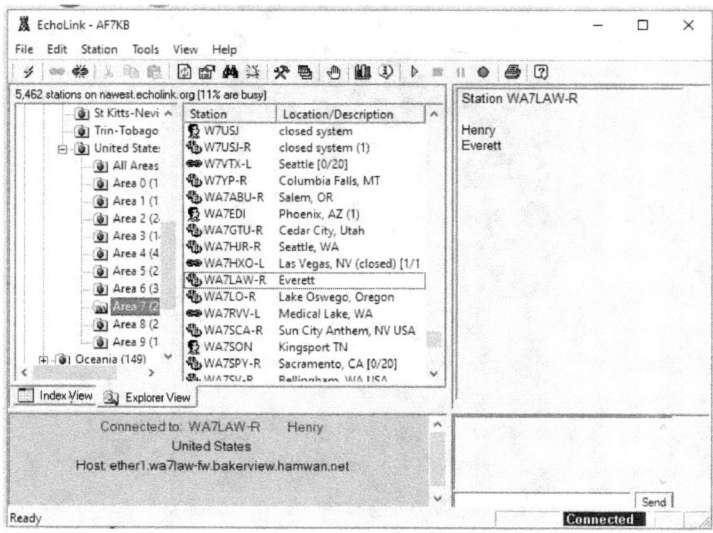

Figure 7.3: EchoLink® Desktop Program

its interface is not as modern looking as the app's. Most of the same functions are present. Since it is the same software that is used to enable EchoLink on a repeater, it's a bit more complicated than the mobile phone app.

You should be aware that EchoLink operates at the speed of the internet, which is not necessarily the speed of light. Especially over long distances, there's a bit of lag; after you start speaking, it takes a little time before your voice appears on the repeater, and the same delay applies to replies from other repeater users. If the user with whom you are speaking is also an EchoLink user, things can get *very* laggy.

Chapter 8

Advanced VHF/UHF

Working repeaters is certainly not the only activity available on the VHF/UHF bands.

Simplex Operation

When you communicate through a repeater, you're "operating in duplex." Put simply, that means you're transmitting on a different frequency than the one on which you're receiving.

It's also possible to operate in "simplex" on VHF/UHF; that means you're transmitting and receiving on the same frequency.

Both the 2-meter band and the 70-cm band have a "simplex calling frequency" and a "simplex frequency." See Table 8.1.

The idea of a calling frequency is that it serves as a "known meeting location." Looking for someone to talk with on simplex? Simple! Tune to 146.520 or 432.100 MHz, say your call sign and "monitoring" and see what happens.

In some areas, there's lots of action on the national simplex calling frequencies, 146.520 and 432.10 MHz. In others, it's quiet as can be. Our area happens to be

Simplex Frequencies	
2-meter Simplex Calling Frequency	146.520 MHz
2-meter Simplex Frequencies	146.400 - 146.580 MHz
70-cm Simplex Calling Frequency	432.100 MHz
70-cm Simplex Frequency	446.000 MHz

Table 8.1: 2-meter and 70-cm Simplex Frequencies

one that keeps 146.520 pretty busy – I haven't heard much at all on 432.10 MHz.

In strict "by the book" operation, you'd establish contact with someone then agree to move your conversation to one of the simplex frequencies; 146.400 to 146.580 or 446.000 MHz. Out here in the real world, I can't recall a time I've done that, nor have I heard someone else do it. Perhaps in a really big city that might be necessary, so don't rule out the possibility, but it's a rare event.

It's certainly worth devoting a memory channel on your handheld or mobile dual-band radio to 146.520 MHz and another to 432.100 MHz.

While it's very similar to talking via repeaters, working simplex is subtly different. When it comes to propagation on simplex, you're much more at the mercy of distance and topography.

Consider the picture of downtown Seattle in Figure 8.1. If you're not familiar with our skyline, that thing that looks like it might have been designed for a science fiction movie is the Seattle Space Needle.

Figure 8.1: Downtown Seattle, WA

On one of our rare clear days, the view from the observation deck near the top of the Space Needle is spectacular, which means there are a lot of places where you can see the Space Needle. If there was a repeater on top of the Space Needle, if you could see the Space Needle, you'd be able to communicate with anyone else who could see it. (Height is might!)

Working simplex, though, that other radio needs to be somewhere where we can "see" each other, so our coverage area is more limited. This sometimes leads to hearing one-sided conversations. There's a ham out on Camano Island who is often on 146.520 on weekends. Camano Island commands a nice view of quite a stretch of Interstate 5, so he's often chatting with hams who are on the road. From my place, I can hear the gentleman on Camano Island great, but I've never heard any of the people on the other side of the conversation. That's just how it goes, sometimes, on simplex.

This dependency on line-of-sight goes a long way to explaining the popularity of the Summits on the Air program. Summits on the Air, far better known as SOTA, is an awards scheme. You can participate in SOTA as an "activator" or as a "chaser." Activators go to the tops of designated summits and try to make contacts from there. Chasers stay at home (or elsewhere) and try to contact the activators. The contacts can be via any legal ham band, including all the HF bands, but 2-meters is one of the favorite bands around here – and we have over 2,000 eligible summits in our state. However, even Nebraska has 15 summits! (Nebraskans won't be surprised to learn they are out in the Panhandle and Sandhills areas.)

If nothing else, SOTA is a great excuse to get out and see some beautiful country, and I suggest that for your first few tries at it, you look at it as exactly that – a fun outdoor adventure that *might* let you make a few contacts. You'll get better at it as you do it more. It's very helpful if you let lots of potential chasers know when you'll be activating a summit, and there's no rule against doing that.

Learn the current rules for participation and find summits here:

https://www.sota.org.uk/Joining-In

There are similar "OTA" programs for islands and parks. Search "IOTA" or "POTA" plus "amateur radio" to get current information on those programs. There's even an only-half-joking "WMPLOTA" event – WalMart Parking Lots on the Air.

While some summits require professional level climbing gear and skills, many others do not. There's a "summit" near our home that is barely recognizable as a hill; it towers 186 feet above sea level! Still, it counts as a summit! Activating a summit, island, or park can be as simple as driving to the summit with your mobile radio, switching it on, setting it to a calling frequency and beginning to operate. Have a notebook and pen ready – even if you're not formally chasing an award, you'll still want to report your contacts to the proper organization. Otherwise, the people you contact – who might be going for an award – won't get credit for the contacts they make with you.

Once you start working locations that require leaving your vehicle, you might begin to have some interest in portable antennas.

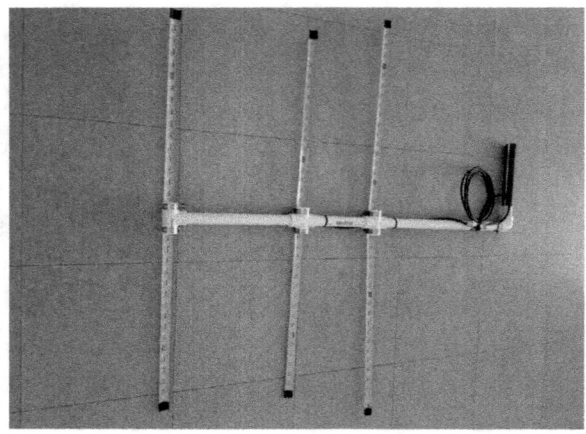

Figure 8.2: Tape Measure Yagi - Courtesy of Michael Hill, WA7MPH

In addition to the portable antennas we've already covered, the "2-meter tape measure Yagi" shown in Figure 8.2 is an easy home-built project. Using it in the field can be as simple as holding your handheld transceiver in one hand and the antenna in the other. The benefit of making it from tape measure material is that it easily folds up for transport. You'll find a video by Michael Marten, KB9VBR, about how to build a tape-measure Yagi here:

https://preview.tinyurl.com/yxt7o7v3

Contesting

There are many, many amateur radio contests every year. A few of them occur on the VHF/UHF bands. Contests are an opportunity to make a lot of contacts and really fine-tune your operating skills.

Generally, VHF/UHF contests are "who can make the most simplex contacts on frequencies above the 10-meter band." As with most contests, there are many ways to score points and many ways to multiply those points.

The first step to participating in a contest is to find the rules and read them. Each contest is unique. Often there are different entry categories; for instance, the June VHF contest has separate competitions for ten different entry categories. You can see them in Table 8.2 on page 129.

When you decide to participate in a contest, it's helpful to have at least a little bit of strategy going. You'll need to choose whether you plan to operate "running" or "search-and-pounce."

Running makes sense if you are a "big gun"; in other words, you have a very

Entry Categories for June VHF Contest	
Single Operator, Low Power	One person performs all transmitting, receiving, spotting, and logging functions as well as equipment and antenna adjustments. Power limits: 6-meter & 2-meter: 200w 1.25-meter & 70-cm: 100w Above 70-cm: 50w
Single Operator, High Power	Power above the limits for Low Power operation.
Single Operator, Portable	Ten (10) W PEP output or less. Portable power source. Portable equipment and antennas. Single Operator Portable stations must operate from a location other than a permanent station location. Single Operator Portable stations may not change locations during the contest period outside of the original 500-meter diameter permitted circle.
Single Operator, 3-band	Restricted to 50, 144 and 432 MHz.
Single Operator, FM Only	All QSOs must be made using Frequency Modulation (FM).
Rover	No more than two operators who move among two or more grid squares during a contest.
Limited Rover	Same as the Rover class but competes using only the lowest four bands available for any given contest.
Unlimited Rover	Same as Rover class but Unlimited Rovers may use more than two operators and are exempt from certain other Rover rules.
Multioperator (Unlimited)	Stations submit logs with more than four bands used.
Limited Multioperator	Stations submit logs with a maximum of four bands used.

Table 8.2: June VHF Contest Rules

powerful station with a big coverage area. A runner finds an empty frequency and calls "CQ contest, CQ contest, CQ contest, this is [call sign] CQ contest", then makes contacts with whomever answers their CQ.

The search-and-pounce operators are tuning up and down the band, listening for those runners' CQ's and are trying to make contact with the runners.

This is a choice of strategy, not a rule of the contest, and you could change strategy from one to the other in the course of the contest. You may also find yourself being a big gun at some point in the contest, and have a bunch of operators trying to contact you – which puts you in the position of being a runner, whether you planned on it or not!

Of course, the overall object of the contest game is to make as many contacts as possible. The *immediate* object of the game is to "make the exchange." That means you and another operator exchange the information called for in the contest rules. Without that exchange matching up in both reports, neither of you will get credit for the contact. In the case of the June VHF contest, the exchange is call signs and maidenhead grid squares. The rules say, "Exchange of signal reports is optional." That being the case, don't expect a signal report! Contests are all about speed, not unnecessary chatter. Here's how a contest exchange might go between me (AF7KB) and KC7YL. KC7YL is running, I'm in search-and-pounce mode.

KC7YL: *CQ Contest this is KC7YL, Kilo, Charlie, seven, Yankee, Lima. Contest!*

AF7KB: *KC7YL AF7KB, Alpha, Foxtrot, seven, Kilo, Bravo.*

KC7YL: *AF7KB, CN82, Charlie, November, eight, two. [My maidenhead grid square is CN82.] QSL? [Did you copy that?]*

AF7KB: *KC7YL, copy. [I understood what you said.] CN88, Charlie November eight eight. [my maidenhead grid square is CN88.] QSL? [Got it?]*

KC7YL: *QSL, AF7KB Thank you! QRZ? [Does anyone else wish to contact me? Y'all come, now, y'hear?]*

Those are the basics of contesting, and it's enough knowledge for you to get in and participate. You should know, though, that there is a lot to the game of contesting, also known as "radiosport." So much, in fact, that the day before the official opening day of Hamvention, the largest US ham convention, is usually devoted to Contest University. Contest University is an all-day series of seminars on all the fine points of successful contesting.

You don't need Contest University to get started, though. You don't need a fancy radio station, either. There's no reason you can't participate in the next contest by standing in your back yard with your handheld, a notebook, and a pen. Later, if you get more serious about contesting, you'll want a real fixed station and, most likely, a horizontally polarized directional antenna, but what's important is to get in there and get started. You'll find lots of resources for you to do just that on the ARRL web site.

My last thought for you on contesting is this; it's a lot more fun to learn these skills from live, experienced people. At our club's field operations, there are always veteran operators ready to coach new people on the skill. Usually, we'll let them watch for a little, then teach them how to run our logging software. That software's pretty simple, but in contesting the log is the whole game, so we take it seriously. (Well – as seriously as we take anything.) Once they've logged some of the veteran operator's contacts, the new ham gets to sit in the operator chair with a script telling them exactly what to say for the exchange. It's all about building skills and confidence.

Digital Voice Modes

Digital voice modes for amateur radio have been around for years but have only recently started to accumulate the "critical mass" of users to become more than just an esoteric activity for a few daring experimenters. It's also only relatively recently that enough digital repeaters have been put on the air to make for a practical communication system.

Digital voice offers three main advantages over analog voice. The first advantage is better signal-to-noise ratio. So long as the radio's signal is being received clearly, the audio on the other end should be very clear and noise free. However, unlike analog, which slowly fades into the noise as distance to the receiver increases, when digital voice drops out it drops *out* – it's gone.

The second advantage is "talk groups." Each digital voice system offers some method of creating a "talk group," a way for groups of users to share a channel at different times without being heard by other users on the channel.

The third advantage is digital linking with other repeaters. That means your talk group could be nationwide – even global! It's like the advanced version of IRLP, the Internet Radio Linking Protocol system that is available on some repeaters.

There are three popular digital voice systems available today. The features they offer are, generally speaking, quite similar. None of the systems is compatible with the other two – they use different modulation and coding schemes.

The three popular digital voice systems are Yaesu System Fusion, D-Star (Digital Smart Technologies for Amateur Radio), and Digital Mobile Radio, known

Figure 8.3: Yaesu FTM-100DR

as DMR.

There are, as of this moment, no "digital only" amateur radios. A few DMR radios are limited to the 70-cm band for both digital and analog modes, but most radios capable of using digital voice modes are full-featured analog dual-banders as well.

- **System Fusion.** Yaesu's entry into digital voice, also known as C4FM for its modulation scheme. Yaesu's advertised primary advantage is a sort of compatibility with analog radios. Yaesu's system automatically detects whether it is receiving an analog or a System Fusion signal. A Yaesu Fusion repeater can also be set up so it can receive System Fusion signals but translate them into analog when it sends them out. The Yaesu FTM-100DR in Figure 8.3 can operate as either an analog dual-band radio or a System Fusion dual-band radio.

- **D-Star.** Originally a closed, proprietary standard. Now, a few homebrewers have made D-Star repeaters and have, it seems, "hacked" the standard. For most people, though, it's still a proprietary standard. D-Star compatible radios are manufactured by Icom and Kenwood. The Kenwood TH-D74 in Figure 8.4 can operate on the D-Star digital system.

- **DMR.** An open standard, DMR is also known as Mototrbo ("Moto turbo"...) because Motorola has used the technology in some of their radio systems. You can see an AnyTone DMR radio in Figure 8.5.

Figure 8.4: Kenwood TH-D74 D-Star Radio

Figure 8.5: AnyTone AT-D868UV (Photo courtesy of Bridgecomsystems.com)

Figure 8.6: OpenSpot 2 Digital Hot Spot

Because System Fusion and D-Star were purpose-built for amateur radio, they're a good deal more "plug 'n' play" than DMR. DMR, however, is significantly less expensive, with Chinese-made DMR radios available starting at less than $160 as of this writing.

https://www.bridgecomsystems.com/collections/anytone-hts/products/anytone-at-d868uv-dual-band-dmr-handheld-radio

I wish I could make a recommendation for you on which system to choose if you decide to explore digital voice. I can't, because I don't know what system is popular in your area. Some parts of the country seem to lean very heavily toward System Fusion, others toward D-Star. Here in the northern part of the Puget Sound area, D-Star repeaters are few and far between, but we have a fairly good selection of System Fusion repeaters. We also have a dedicated group that is creating a series of DMR repeaters to serve the I-5 corridor, and that project is nearing completion. In other parts of the state, DMR is almost non-existent – so it very much depends on what's happening in your local area.

The way to start learning about what digital voice system is popular in your area is with a search of repeaterbook.com.

Digital Hot Spots

No digital repeaters in range of your house – or, at least, none that are connected to the internet? Digital hot spots are your gateway to the world.

Figure 8.6 shows a digital hot spot.

A digital hot spot is a lot like a mini-repeater with an internet connection. The one in Figure 8.6 contains a wi-fi router and a UHF ham radio transceiver, plus a micro-computer. It connects to your wi-fi internet connection and you can talk with it using a UHF handheld ham transceiver. In effect, it's your very own private repeater.

OpenSpot isn't limited to a single digital format, either; it is compatible with DMR, D-STAR, System Fusion/C4FM, and more.

Once you're connected through a hot spot such as OpenSpot, you can connect with talk groups all over the world – you might be connected to 100 repeaters at once, with nearly global coverage.

If you decide to play in the DMR digital voice realm, I'd suggest you take a look at the offerings of BridgeCom Systems.

http://bridgecomsystems.com

Ron Kochanowicz, KC0QVT, the President of BridgeCom, and the rest of the crew seem to be an endless fountain of useful knowledge, and their educational systems are excellent.

Part III

Getting Started in HF

Chapter 9

Building Your First HF Station

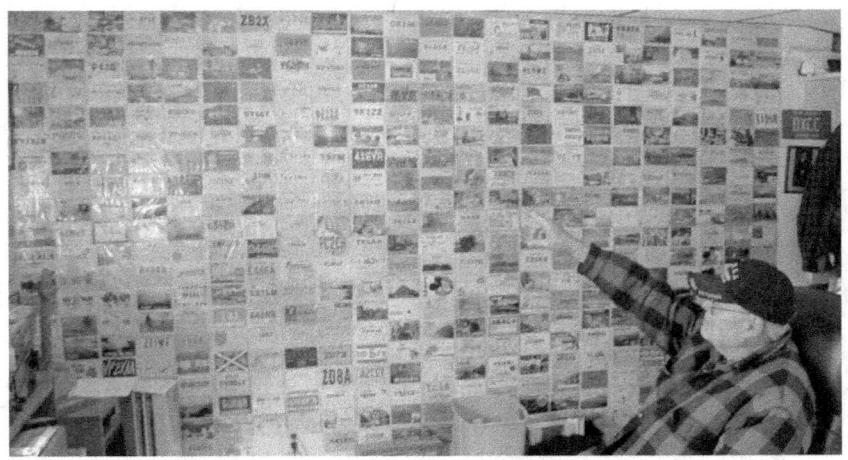

Figure 9.1: This Could be You!

I think it's safe to say that there are no two amateur radio HF stations that are exactly alike. Every one of them is unique, and created in response to the ham's operating goals, the operating site, and the ham's budget. Sometimes they're also the product of what I'll call "ham lore." That's the mis-information that seems to have a life of its own as it echoes around the ham community!

Some quick basics. When we talk about HF, we're talking about the High Frequency bands; those below the 6-meter band. Primarily, those are the 160-meter through 10-meter bands. There are a couple of more ham bands way down there below 160-meters, but for the moment those are primarily experimental, and you won't find them on commercial HF transceivers. And yes, I know I should speak of the 160-meter band as MF, Medium Frequency, but, at least in casual hamspeak, 160-meters gets lumped into the HF bands.

The HF bands are the worldwide communication bands – they propagate over long distances, mostly with the help of the ionosphere. The condition of the ionosphere varies with the condition of the Sun. For us, the best indicator of the condition of the Sun is the presence or absence of sunspots. More sunspots usually equals better propagation. Sunspots vary in their frequency on a 22-year cycle, with peaks every 11 years.

You've probably heard we're in one of the valleys of the sunspot cycle, and you may have even heard that the HF bands ARE DEAD!

First, some good news. No matter what you may have heard, THE BANDS ARE NOT DEAD – they just smell a little funny!

It's true that we're in the bottom of the sunspot cycle, and it's also true that HF propagation is certainly not what it will be when the sunspots start appearing in greater numbers. However, there haven't been many days at my station – which is, by the way, nothing special – when I couldn't make some contacts somehow. So far as I know, the ARRL continues to send out new DX awards. Yes, propagation conditions have been and, at least for a while, will continue to be challenging. I won't say the bands are alive and well, but they're alive.

Figure 9.2 shows a recent propagation forecast. You can see, this wasn't a great day for propagation on any band above 40 meters.

Supposedly the only good propagation available was on the 80- and 40-meter bands, and then only at night. And yet ……

Figure 9.3 is a picture I snapped of the DXMaps.com site that same morning. We're looking at contacts reported on the 20-meter band.

So, there's plenty of fun to be had on the HF bands, and it doesn't take a 100-foot tower festooned with exotic and expensive antennas connected to the world's greatest radio to get in on the HF action.

Getting on the HF bands vastly expands your ham radio world – both geographically and operationally.

On HF, there's a wider variety of activities. Obviously, there's CW and SSB phone, and lots of it. There's also quite a selection of HF digital modes, some of which have become enormously popular, at least for the moment, in response to propagation conditions. . You can have interesting conversations with people all over the world or you can go for awards like Worked All States or even DX Century Club, which is awarded to hams with verified contacts with 100 "DX entities"; basically, 100 different countries. There are lots of contests of all sorts.

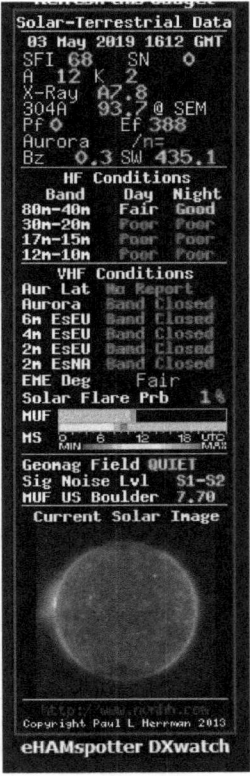

Figure 9.2: Propagation Forecast

If you enjoy nets, there are plenty of them to choose from across the bands.

Each band tends to take on its own character. That's partly because of how the frequencies are affected by propagation conditions, and partly just by the semi-random assortment of folks who have chosen to hang out on those bands.

So, what is it that you want to do on the HF bands?

"Well, Mr. Burnette, I don't know! That's why I bought your book!"

I get it! It's a big smorgasbord, and the truth is you won't know what you enjoy and what you don't enjoy until you get in and try a few of these things. You just can't tell from the outside. When I first heard about the digital mode called FT8 I thought it was absolutely not for me. It's very limited in what messages it can send – it's like a chat program that can only say CQ, my call sign, your call sign, a signal report, and 73! Dumb! Then I fired it up on an old laptop running through my disco-era transceiver into a very inadequate antenna and it turned out to be a lot of fun.

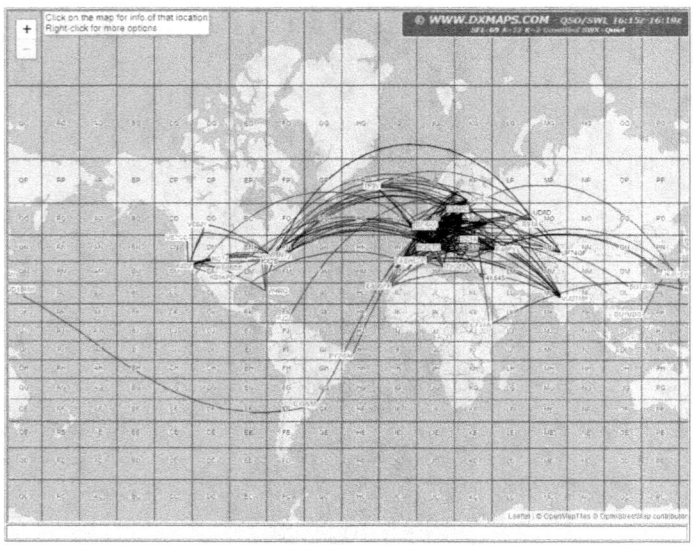

Figure 9.3: DXMaps.com QSO Map

Planning Your HF Station

Since you're starting off on your HF journey, let's think in terms of a station for you that will let you experiment with a lot of HF modes and bands so you can find out what lights you up. Let's add to that, too, that we do this at a reasonably modest level of expense and inconvenience, so you're not in over your head before you even get started. When I say modest expense, I mean by the time you finish building your very own radio station capable of global communications, you'll have a lot less into it than the two knuckleheads in Figure 9.4 each have into their bicycles and bike gear.

That's our son and son-in-law on their high-tech bicycles – we *know* what they have into their hobby!

Let's specify that this will be a station capable of operating multiple bands, and since it's well within reach, let's aim for everything from 80-meters through 10-meters. 160-meters is sort of its own animal, especially when it comes to antennas, so we'll set that one aside for later.

Let's go with modest power – say 100 watts. QRP – low power operation – is a great branch of the hobby, but probably not the greatest starting point. QRO, high power operation, with a big amplifier, can come later, when you have your antenna and operating skills sorted out.

Figure 9.4: A Lot of Money in the Form of Bicycles and Gear

We'll talk more about transceivers, but with those specifications, just about any mainstream all-mode all-band HF transceiver made in the last two decades will do the job.

We covered the basics of choosing a spot for your shack in the dual-band fixed station chapter, back on page 78. The way I put it is, "Those are the same for HF, only more so!" In other words, you'll need more space for gear – including, most likely, at least a laptop computer – less noise, etc. Fortunately, today's HF radios are considerably more compact than HF radios were back in the "boat anchor" days. Our "shack" is a table in our breakfast nook, and it?s adequate. (It's also close to the coffee maker – a definite plus!)

You'll do well to focus on the electromagnetic noise aspect of your shack, as well. Because HF receivers are so sensitive, the constant enemy of HF is noise. The less noise your immediate environment contributes, the better. Our location is plagued with noise from overhead LED lights in our kitchen; we do a considerable amount of radioing in semi-darkness. Big screen TV?s can be very noisy, and even some modern washers and dryers have digital controllers that make quite an electromagnetic racket.

Speaking of noise, please be sure to read the chapter on Grounding the Amateur Station, starting on page 91. It applies to VHF/UHF stations and HF stations as do all the mobile and fixed station installation and construction techniques in the VHF/UHF section.

HF Antennas

Let's start with what can be the most physically challenging item; the antenna. Any experienced HF operator will tell you; the antenna is the single most important piece of equipment in your station.

To some extent, your choice of HF antenna will inevitably be dictated by the space available to install it and other issues related to that space.

If you have a few acres of land with some very tall trees, no neighbors around for miles, no homeowner's association to deal with, and an unlimited budget, you have a lot more possibilities than someone on a tight budget in a rented apartment with no land and a landlord who is rabidly anti-antenna.

If you have the space, and some tall trees or something else tall to hang a long-wire antenna, that's what I'd recommend for your starter. You can buy commercial versions of several popular designs. If you're even a little bit handy, they're simple enough to construct, and while I won't say they're effortless to put up, they aren't major construction projects. Long-wires strung on trees are very stealthy, too – you usually have to look hard to see them at all.

There are three basic configurations for long-wire antennas. Flat-tops, slopers, and inverted-v's. The flat-topper (Figure 9.5) looks like a lot of the antenna

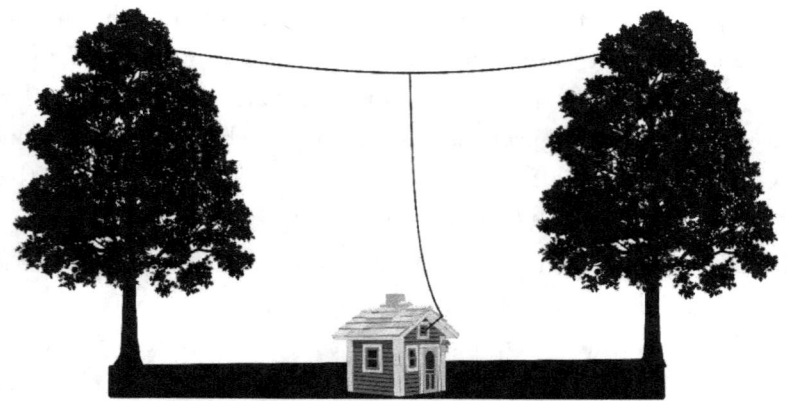

Figure 9.5: Flat-top Antenna

pictures in your license manuals look – a more-or-less horizontal wire. In this one the feed point's in the center, but it could also be at either end or even at a place somewhere between the center and the end. Moving the feed point changes the impedance and performance of the antenna, so do a little research into the various sorts of long-wire antennas.

A sloper runs from a lower spot to a higher spot – usually that works out to be from the eaves of your house up to some tall object. Again, it could be center-fed, or off-center fed, but I think most slopers are end-fed.

Figure 9.6: Sloper Antenna

An inverted V runs from a lower spot to a higher spot, and back down to a lower spot. There are end-fed, off-center fed, and center-fed inverted V designs, too.

Figure 9.7: Inverted V Antenna

There are a few multi-band long-wire designs to choose from. One popular one is the G5RV, and that would be my first recommendation for you if you have the space and a place to hang it.

Figure 9.8: Ready-made G5RV Antenna

You can buy these ready-made from several sources – the one in Figure 9.8 happens to be from MFJ – or you can make one yourself with some wire, some insulators, a length of ladder line, also known as window line, and a choke balun. You do need an antenna tuner to make a G5RV work, and most likely an external tuner – most tuners built into transceivers won't handle the range of SWR (Standing Wave Ratio) from a G5RV. (G5RV is the call sign of the English ham who invented the antenna.)

A full-size G5RV is 102 feet long. The center needs to be at least 35 feet off the ground. It can be a flat-topper, sloper, or inverted V. If you set it up as a sloper or inverted V, the lowest parts need to be up at least 25 feet. As with just about any antenna, higher is better, so if you can get that center-point up 100 feet, go for it. It will make the radiation pattern better and be a lot easier to tune.

One more long wire for you that's a terrific antenna; if you have a big chunk of land with several tall objects near the perimeter, you might consider a horizontal loop.

You'll need three or four – or more -- good sized – say, 40 feet tall or more – trees or other tall things on your property, that are spaced a healthy distance apart.

The length of the antenna is one wavelength of the lowest frequency for which you are building it. If you have room to string out about 160 meters -- 550 feet -- of wire, you can have an antenna that will work 160 meters through 6 meters.

Figure 9.9: Horizontal Loop Antenna

We're aiming for 80 meters and up, so you'll need to string about 275 feet of wire to create a horizontal loop antenna. The horizonal loop isn't terribly picky about its exact shape – square, rectangle, even triangle shapes work. The idea is to enclose as much area as possible with the wire. It isn't picky about height, either. Higher is better, but there are reports of some installed as low as 10 or 20 feet high and, by all reports, they work.

Are there more long-wire antenna designs? Absolutely. Whole books of them. But, given the parameters we've chosen, is it worth spending days researching them? Meh … maybe. By all means, learn about them, but you probably won't find a fantastically more efficient multi-band long-wire antenna than what we've covered. What's important at this stage is to get on the air and start learning and experimenting – you can optimize your setup later when you really know where you want to focus.

The basics of putting up long-wires are pretty simple. Your objective is simply to safely get a hunk or two of wire up in the air as far as you can, separated from the supports by insulators. The ends should be supported by non-conductive material – usually rope. You'll want to rig the ends on some sort of system that lets you raise and lower the antenna to get it tuned to the proper length.

If you're using a tree or two, there are a number of ways to get those support ropes up in the upper reaches of the tree – none of them involves imagining you have skills in climbing trees. My advice is to leave that to cats and to the people who get paid the big bucks to do that kind of work.

If you're handy with a fishing rod, you can cast a lead weight over a tall branch,

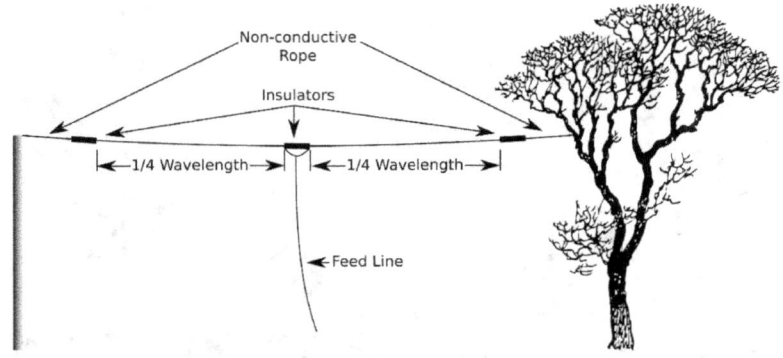

Figure 9.10: Hanging a Long-Wire Antenna

use the fishing line to pull up a stouter rope, then use the stouter rope to support the antenna. People have been known to tie some fishing line to a bow and arrow and get the job done – this is probably a good time to say, SAFETY FIRST ...

One favorite is an air powered gun like the one in Figure 9.11, advertised regularly in the back pages of QST Magazine, that fires a spherical fishing weight. This is a device that really shouldn't be nearly as much fun as it is.

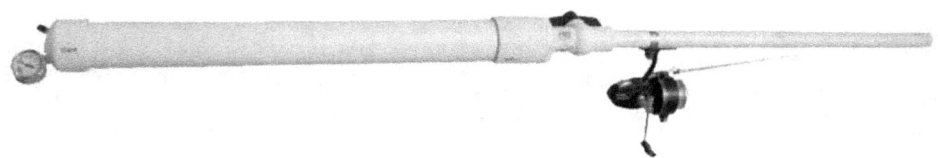

Figure 9.11: Air Boss Antenna Launcher. Photo courtesy of Bill Olah, KR4LO. Used by permission.

If you're putting up a very temporary installation, you can use something as crude as nylon packing twine for your support rope. For something more permanent, you'll want some Dacron rope that's rated for exterior use. Be sure to hang the antenna so there's some slack – you don't want it taut when the wind starts to blow!

A lot of times when I tell people who are new to this about how to put up a long-wire antenna they kind of give me this look, like, "It can't be that simple." It really can, though! Our club puts together a field operation every year for 7QP, and one of our favorite field antennas is our fan dipole for 80, 40 and 20 meters. With some helpers, it takes less than an hour to set up the whole thing from the moment the antenna comes out of the box.

A little hint if you're in a ham club – you can accomplish a huge amount of antenna building in a very short time with a couple of 16" pizzas and a cooler full of beverages.

Nothing tall around your house? I'm in that same boat – there's nothing even close to high enough to hang a decent wire antenna at my place. However, I do have a fair sized back yard, so I've put up a vertical.

It's possible to homebrew a vertical antenna, but there's no getting around the fact that they present more difficult engineering problems to solve – and that's both electronic engineering and civil engineering. After all they have to stand up – and stand up even in stiff winds. By the time you scare up all the various parts you need to assemble a multi-band vertical, in most cases you might very well save money, time, and frustration by simply ordering a commercially made one.

HF verticals are just about universally quarter-wave antennas with loading coils – a true half-wave 80-meter vertical would tower some 132 feet in the air. That's a lot of aluminum to make self-supporting, even with guy wires. Quarter-wave antennas are a LOT more practical, but because they're quarter-wave antennas they typically require some sort of counterpoise or ground radials to work.

Radials are simply wires stretching out from the base of the antenna. A counterpoise is typically elevated above the ground. Counterpoises must be a precise length in order to function properly at the design frequency. Ground radials are not length sensitive; basically, "longer" is better, up to the length of the antenna, and "more" is better. Commercial AM stations usually have one radial every three-degrees, for a total of 120 radials, each the same length as the antenna is high, however that's because their design is focused on creating lots of ground wave and not much sky wave. For our purposes, we don't need 120 radials around our antenna; we're not going to create much in the way of ground waves on bands above 80-meters, no matter how much wire we lay down.

Figure 9.12 is a 43-foot Zerofive. Basically, it's a 43-foot-high aluminum whip antenna with ground radials all around the base. It covers 10 through 160 meters.

Figure 9.13 is a HyGain DX-88. It's 25-feet tall and covers 80 through 10 meters. I just had a chance to see one of these in action, and to compare it on several bands with a dipole and a tri-bander Yagi on a tall tower. The DX-88 was a little noisier, but certainly held its own. At times, it was outperforming the dipole. It was working off about 20 ground radials that were about 25-feet long. This was a field operation, so we just laid them on top of the ground – if you were putting this in your back yard, you could bury the radial wires or just use turf

Figure 9.12: 43-foot Zerofive Vertical Antenna (Picture courtesy of Floyd Larck, KK3Q)

staples to hold them down while your grass grew up around them.

Figure 9.14 is a radial plate that goes at the base of a vertical antenna to connect all those radial wires. We own one of these and met the ham who makes and sells them – he's a machinist out of upstate New York and sells on e-bay.

That DX-88 we were using was ground-mounted – other vertical HF antennas can be roof mounted.

There are at least a couple of multi-band verticals that advertise that they don't require radials but even those are said to work better with some sort of radial system installed – even some radials strung out across your roof seem to improve them.

That brings us to the topic of stealth antennas. There are still plenty of zoning ordinances and Homeowners' Association regulations out there that forbid or drastically limit what sorts of antennas – if any – we can put on our property.

However, there is still hope. There are lots and lots of ideas on the web for stealth antennas – here are a few of the more common solutions.

Most towns and HOA's don't forbid flagpoles – after all, that would be downright unpatriotic, right? Well, look very closely at the handsome flagpole in Figure 9.15 and you'll see some suspicious lumps --- loading coils and traps! That's Zerofive Antenna's 30-foot flagpole antenna that covers 80 through 10 meters.

If you have some long lengths of plastic rain gutter around your home, you can

Figure 9.13: HyGain DX-88

Figure 9.14: Radial Plate

Figure 9.15: Zerofive Flagpole Antenna

sneak an end-fed half-wave wire antenna into the gutters, even if it has to turn a corner or two. If you have aluminum gutters – well, I know for a fact that there's at least one set of rain gutters in my neighborhood radiating down to 40-meters. It may take a little ingenuity, but it absolutely can be done – I've seen it work. If your gutters circle all or most of the house, you have the makings of a horizontal loop.

You say you live in a high-rise apartment or condo? If there's a balcony, you can stick a Hamstick or a screwdriver antenna out there and dangle some 14-gauge wire down the side of the building for a counterpoise. You're not going to be a Big Gun contest station with that set-up, but you'll be operating!

Do you have a big attic? You can stick a long-wire up there. It's also possible to create a compact long-wire antenna from a couple of Slinky toys! Check out the link below for one ham's version.

https://www.nonstopsystems.com/radio/frank_radio_antenna.htm

An attic antenna won't be optimum, but it can work, so long as everything is wood up there. If you have a metal roof – you're out of luck on putting up an attic antenna. If you go the attic antenna route, please remember, your antenna

Figure 9.16: Magnetic Loop Antenna

is carrying some very high voltage, especially at the ends, so observe proper precautions. The League publishes a whole book of stealth antennas – well worth the investment if it saves your house from burning down!

Or, there's the magnetic loop antenna.

The one in Figure 9.16 is from Alpha Antennas, and there are other companies that make them, as well. They're pretty easy to build for yourself – right up to the moment you try to find that heavy-duty high voltage capacitor the antenna must have to work as a transmitting antenna. You can see it in the picture; it's the white box at the bottom of the loop. The rest of the antenna is just a couple of lengths of coax. Only the shields of the coax are conducting – the center-conductor is just along for the ride. Your feed line attaches to the small loop, that induces a signal in the large loop, you tune the big capacitor for minimum SWR, and you're radiating. Kind of inconvenient – you have to retune them every time you change frequency. They're not the most efficient antennas in the world, either, but they work, as countless DXpeditions can attest. Magnetic loops will work effectively just 36 inches off the ground. You can mount one on a tripod, stick it out on your patio, and operate world-wide without the HOA ever catching on – and they wouldn't have a basis for a complaint, anyway, because that's not a permanent installation.

One more possibility – and this one is for you if every possibility I've mentioned has been, for whatever reason, not workable. If you have internet access and a couple of pieces of the right equipment and/or software, you can operate a rig remotely – such as a club rig.

A friend of mine has been a ham since 1951. Yep, 1951. He recently moved into a living situation where there simply is not the room for any sort of antenna. No problem – he became a member of an amateur radio club that has a remote-controllable station and operates that way.

A Google search for "remote ham radio station" will show you some possibilities. One very useful site that lists lots of remote-controllable stations is remotehams.com.

http://remotehams.com/online.html

On the deluxe side of things, for as little as $99 a year plus a "per minute of use" fee, you can have access to ten 500-watt stations driving big directional antennas on tall towers through the Remote Ham Radio company. The stations are placed all over the United States. Boost your membership to their "PremiumDX" package for $999 per year (plus "per minute" charges) and you'll have access to 21 stations placed around the country. That's access to, conservatively, hundreds of thousands of dollars worth of "megastations".

http://www.remotehamradio.com/

I'm not going to go into all the intricacies of setting up a remote-control system, because there are a lot of ways to skin that particular cat. Obviously, the radio being controlled has to have computer control capability. One system works directly from your computer by running software like the Ham Radio Deluxe remote-control server – you end up operating from a screen that simulates the front face of a transceiver; it looks like Figure 9.17.

That software talks to an interface of some sort at the other end, and the interface talks to the remote-controlled radio.

HF Feed Lines

Before we go radio shopping let's talk briefly about your feed line. Now, what I'm about to say borders on ham-lore heresy, so shield your delicate eyes if you're going to collapse or something, but here we go – it's perfectly possible to have a working feed line of that RG6, 75-ohm coaxial cable you can buy at Lowe's or Home Depot for about 30 bucks for a 100-foot roll. GASP! But, but, but – that means an automatic 1.5:1 SWR! HORRORS! So what? 1.5:1 SWR translates to 0.18 dB of loss – about 4% of your power. If the rest of your antenna system is well matched to 75-ohms – which will take some work – you're probably getting more signal to your antenna than half the HF operators in the world.

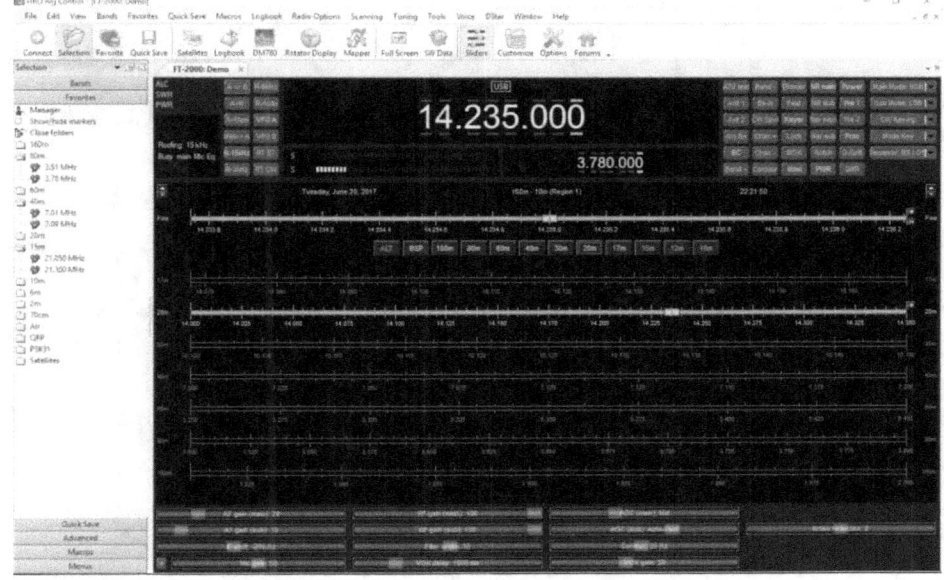

Figure 9.17: Ham Radio Deluxe Remote Control Screen

However, by the time you buy some conduit in which to install that feed line, the connectors, and the matching gear you'll need at the antenna end, you'll be into the deal about $70 or $80.

The good folks at DX Engineering will quite happily make you a 100-foot length of their "400 Max" low-loss, double-shielded coax – complete with professionally installed PL-259's on either end – for $119.99 with free shipping. It's double shielded, direct buriable – no conduit needed, and only loses 0.8 dB per 100 ft at 30 MHz. If you remember to order some waterproofing tape for your connectors while you're ordering that coax you'll get free shipping on that, too. I imagine GigaParts, HRO (Ham Radio Outlet) and MFJ will just as cheerfully make you similar deals on similar products. If you have a shorter run than 100 feet, you get it handled for even less!

There's the small matter of how that feed line gets into your ham shack. Of course, you can go with the time-honored, "drill a hole in the wall and run the coax into a junction box with a PL-238 bulkhead connector on the front. " Some folks don't feel comfortable drilling and sawing holes in their walls, so for them, one clever solution, from MFJ, is shown in Figure 9.18.

What you see in the picture is the exterior side of the panel but the interior looks the same.

Figure 9.18: MFJ Coax Window Feed-Through

That panel comes mounted in a cedar board that you cut to size to fit in your window – the board has weather-stripping and everything. About as fast as you can measure and saw, you have a weatherproof, bug-proof feed-through for your feed line, and you didn't have to drill any holes in the wall. That particular one costs about $70 – if you count up the $5 connectors on there, you probably can't build it yourself for that. Notice, it even includes a feed-through for your ground. Slick!

And that brings us to the subject of your HF radio.

Choosing Your First HF Transceiver

Modern HF radios are certainly amazing. Look at Figure 9.19....

Figure 9.19: ICOM IC-7851

That is the Icom IC-7851. It's a beautiful piece of technology, and it costs about $12,500 unless you catch it on sale. Look at all the knobs and buttons on that thing! I count 97 of them, but I might have gotten lost in the middle of the count.

You don't need that radio. That's a Steinway and you still need to learn how to play Chopsticks.

Let me gently remind you that many years ago, hams all over the world were communicating with each other just fine with radios like the venerable Collins

Figure 9.20: Collins KWM-2

KWM-2, shown in Figure 9.20 – a radio with a total of ten knobs and switches including the on/off switch and the three-position (three!) meter function switch.

I'm not advising you to run out to the swap meet and find a KWM-2 for your shack – but I am pointing out that you honestly don't *need* every single modern doodad, gizmo, and whatchamacallit to get started and operate quite successfully on the HF bands.

Most of the major makers make good, solid entry-level HF radios that are more than adequate for your present needs.

You want a radio more like the one in Figure 9.21.

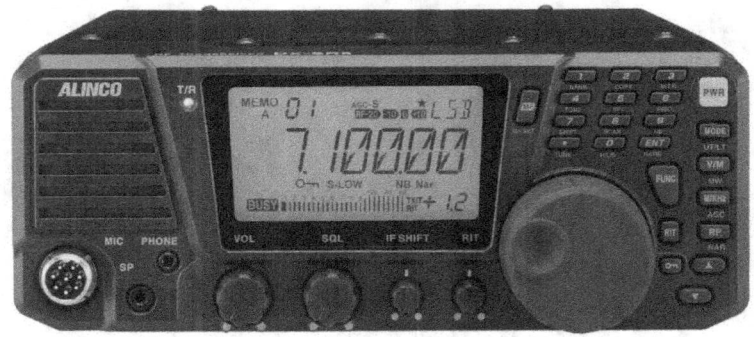

Figure 9.21: Alinco DX-SR8T

That's an Alinco DX-SR8T. List price about $500 brand-spankin'-new. It has a tuning knob, a volume knob, squelch, RIT, and IF shift, and covers 160 through 10 meters. We just had one operating at 7QP and it chugged right along and did

Figure 9.22: Yaesu FT-891

its job. In fact, it kept going when one of the more spendy radios pooped out halfway through the afternoon!

For roughly another $150 you can step up to Yaesu's FT-891. It covers 160 through 6 meters. It's a very compact radio, so many functions require stepping through the unit's menus, but the payoff is small size and a lot of features.

If your preference runs to ICOM, they offer the IC-718 at almost exactly the same price as the Yaesu FT-891, about $600.

Figure 9.23: ICOM IC-718

By stepping up a little you can get a *lot* more radio with ICOM's IC-7300, Figure 9.24, with a list price of about $1300. Our club owns and runs four of these for our major club events. Without getting too deep into the technical details, let's

Figure 9.24: ICOM IC-7300

just say the 7300 offers a lot of advantages because it is, for practical purposes, a Software Defined Radio.

If you're up for a little risk, there's always the used market for ham radios.

Figure 9.25: Used Kenwood TS-520S

If you choose to go the used route, I'd advise you to stay as local as possible, or at least face-to-face at a hamfest. I've heard way too many horror stories about e-bay ham radio transactions to recommend those to you. You want to see that thing at least light up and make noise – it's even better if you can test the transmit side into a dummy load with a wattmeter. Then you know you're buying a radio, not a project.

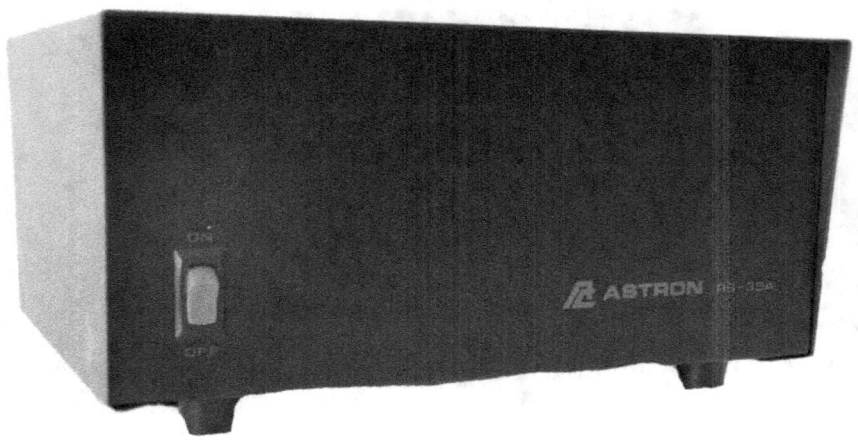

Figure 9.26: Astron RS-35A

It is useful to have a sense of humor if you're going used radio shopping. The note on the swap meet Kenwood TS-520S in Figure 9.25 says, "Everything seems to work ok. No CW filter. 120 watts output on 75 meters. The transmit and receive frequencies are off each other by about 100 Hz. This offset is solved by always using the RIT to align the frequencies." (We'll cover the mysteries of the RIT in the next chapter.)

That used radio has some ...let's call them "quirks", but it's a fine old radio and a steal for what you could have paid for it by the end of that swap meet.

Power Supply

You'll need a power supply for your radio, and one that's capable of delivering all the power the radio needs.

If anything, the demands on an HF power supply are steeper than those on a typical VHF/UHF fixed station power supply. The reason is that the enemy of HF is noise, and anything you can do to reduce noise on your received and transmitted signals will add up to more contacts. Having an electrically quiet power supply is key to having minimum noise in your signals.

Personally, I'm a fan of center-tap type power supplies, like the Astron 35-amp model in Figure 9.26.

You won't be bringing that power supply home balanced on the handlebars of your bicycle – it has a *big* transformer in it and it weighs around 20 pounds, but mine has been running since the 1980's and hasn't missed a beat.

There are also switch-mode type power supplies, such as the MFJ-4125, Figure 9.27.

Figure 9.27: MFJ-4125 Switch-mode Power Supply

It's lighter and smaller than the Astron. It's only rated for 25 amps, not 35, so it's just barely adequate for most 100-watt HF radios, but it's adequate. It's also considerably less expensive than the Astron, at $89.95 versus about $220.00.

Antenna Tuner

The last major item that will probably be on your list – depending on the antenna choice you make – is an antenna tuner.

Figure 9.28 is the MFJ-949E, which MFJ claims is the most popular tuner in the world. It has lots of range, plenty of power handling capacity – 300 watts – and a new one will set you back about $200. For about $50 less, there's the MFJ Travel Tuner, the MFJ-945E, shown in Figure 9.29.

That one also is rated at 300 watts maximum.

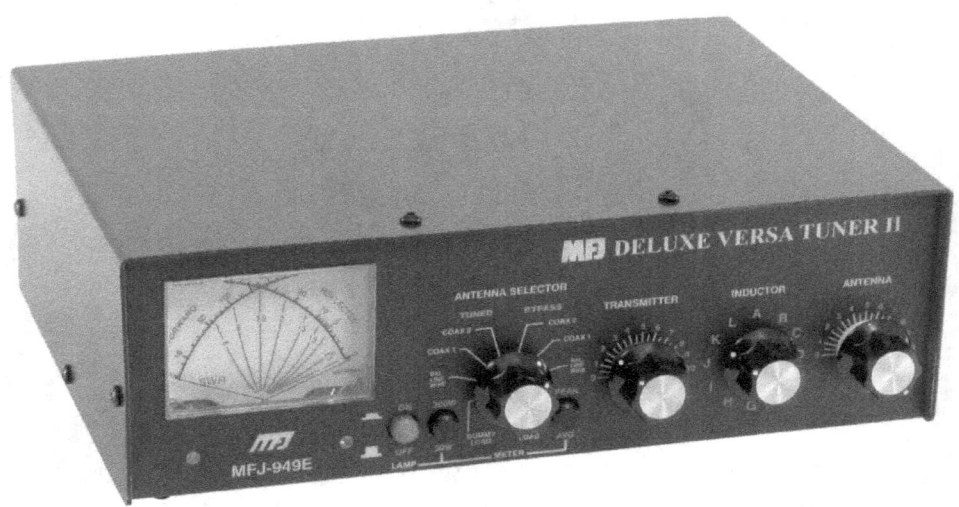

Figure 9.28: MFJ-949E Versa Tuner II

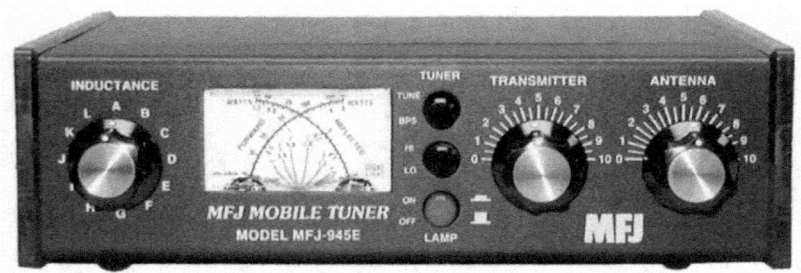

Figure 9.29: MFJ-945E

Chapter 10

HF Transceiver Controls

I can still remember being about 13 years old and seeing my friend Brent's ham radio station. I knew right away this guy had to be some kind of genius to be able to operate what was obviously a *very* complicated machine that required abilities far beyond the reach of mere mortals!

Figure 10.1 shows the transmitter and receiver he had. They look a lot less intimidating to me now, but I want you to know that if you've felt like these things are pretty complicated, you're not alone. More important than that, though; it's possible to learn.

 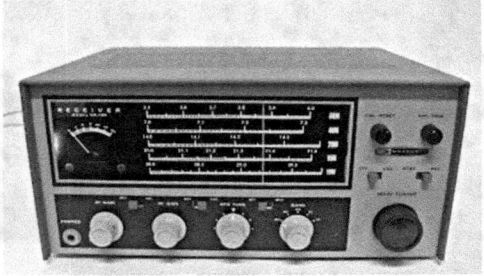

Figure 10.1: Heathkit DX–40 Transmitter (L) and Heathkit HR-10 Receiver (R) DX–40 Photo Courtesy of AB2RA, Janis Carson. HR-10 Photo Courtesy of K6JCA, Jeff Anderson.

HF radios do have a lot of controls even when compared with dual-band amateur radios, much less when compared to consumer broadcast radios, and Brent was (and still is) a very bright guy, but HF operation does not require genius. Thank goodness!

Why do we have so many knobs and buttons?

One reason we have all those controls is because we expect our radios to do a lot of different jobs, from sending Morse code to receiving slow-scan television pictures.

Most of the controls you see on an HF transceiver are connected to the receiver side of things. Even on the fanciest transceivers, there aren't many controls for the transmitter side; you can adjust frequency, mode, and power, and maybe some sort of rather primitive processing for the transmitted audio. That's about it. On the receiver side, though, radios can have all sorts of features that help us improve a very faint, noisy signal just enough to be able to understand it. Those ninety-seven buttons on the ICOM IC-7825 are almost all devoted to various filters and processors on the receiver side.

As complex as that radio is, its controls are not fundamentally different from those on the entry level radios. Once you know how to operate one HF transceiver, you'll be able to pick up how to operate another fairly quickly. Let's take a tour of an HF radio.

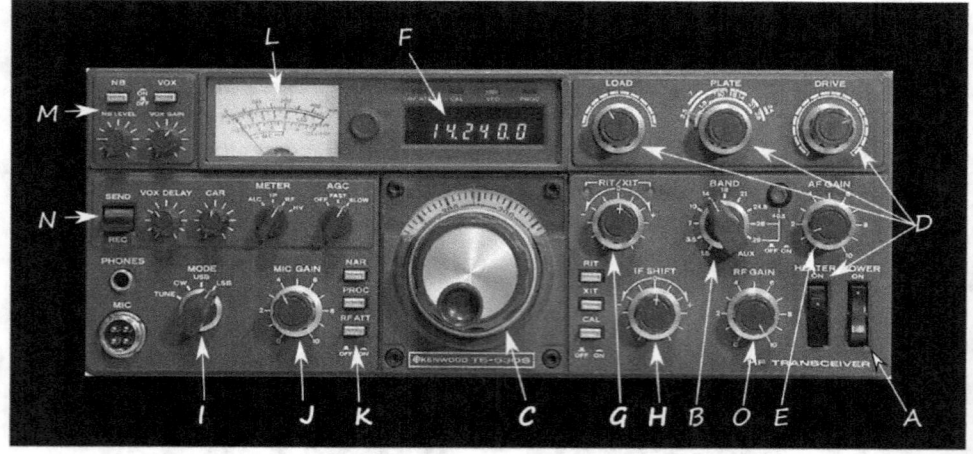

Figure 10.2: Kenwood TS-530s

That Kenwood TS-530s in Figure 10.2, page 164, is an oldie (but a goodie!) They were immensely popular in the 60's, and you'll still find used TS-530's for sale at most hamfests. You'll find many of the controls on that radio on every HF transceiver made today. I'm not recommending that radio to you, it's just useful for this purpose because everything is a knob or button instead of a menu item on a screen.

A) That's the power switch.

B) The Band switch lets you select your transmit/receive band. It is set on the 20-meter, or 14.0 MHz band. If you look closely, there's a small button to push to select "+0.5" MHz, which only affects the 10-meter bands.

C) The VFO knob is how you select a frequency within the transmit/receive band you chose with the band switch.

D) This transceiver is a "hybrid" transceiver. It's all-transistor except for the final amplifier, which used vacuum tubes. At the time this was made, there just weren't affordable transistors that would amplify 100 watts of RF. The Load, Plate, Drive, and Heater controls all have to do with those vacuum tubes. You won't find those on modern transceivers, though you might find them on an RF power amplifier. Should you buy a vintage unit like this one, you can find a manual for it on line, which will demystify those controls for you.

E) That knob marked "AF Gain" (Audio Frequency Gain) is the volume control.

F) Frequency read-out. You'll notice there's an analog read-out just below the digital one. I suspect that's because at the time this radio came out, those digital read-outs were newfangled contraptions.

G) The RIT/XIT control is one you'll use a lot on SSB. Sometimes known as the "clarifier", RIT stands for Receiver Incremental Tuning and XIT stands for Transmitter Incremental Tuning. There are buttons just below the knob to select what the knob will control; RIT, XIT, both, or a calibration setting. The one you'll probably use most is RIT. It lets you adjust your receive frequency over a narrow range without changing your transmit frequency. It lets you tune the sound of SSB to sound like normal human speech instead of Minnie Mouse or Darth Vader.

H) IF Shift is a type of filtering for excluding interfering signals that are very close to the carrier frequency you're trying to hear. IF shift filtering happens in the Intermediate Frequency stage. Its purpose is to remove a nearby interfering signal before it gets through the IF stage, because after that it's pretty much impossible to filter out.

Imagine this scenario. You're on the 15-meter band, trying to hear someone on 21.115 MHz; there's an interfering station on a frequency just below where you're listening, and their signal is overlapping into the signal you're trying to hear. See Figure 10.3.

Tune the radio a little higher? You would lose some (or all) of the signal strength, and on SSB you'd shift the pitch of the voice you were listening to. That really doesn't work.

How about using the band-pass filter? That narrows (or widens) the frequency

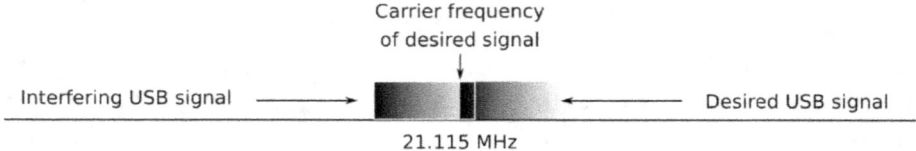

Figure 10.3: Interfering Signal Overlapping Desired Signal

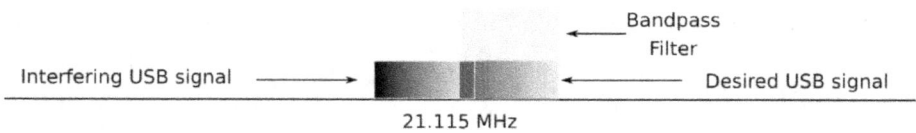

Figure 10.4: Normal SSB Band-Pass Filtering

spectrum you're listening to. That USB signal we're trying to hear is about 3 kHz wide, and it looks like that other signal is overlapping our desired signal by at least 600 Hz. We have our band-pass filter set to something like you see in Figure 10.4.

Since both "sides" of a band-pass filter move together, that means we need to narrow our band-pass by 1200 Hz, as in Figure 10.5.

That doesn't leave us with much signal, and intelligibility on SSB is challenging as it is. Narrowing the band-pass filter is worth a try, but wouldn't it be great if we could move just the left-hand side of that band-pass?

We can, if we have IF shift on our receiver. That's the trick of IF shift. The trick is accomplished in the IF, Intermediate Frequency, section of the radio, but don't worry about that – for now, just get this concept of moving one side of the pass band, so you can end up with the situation shown in Figure 10.6.

Compare that illustration to the one in Figure 10.5 and you can see we have preserved a lot more of the signal we're trying to hear. By leaving the radio tuned exactly to the received station and using the IF shift, we keep the same signal strength but listen to a slightly different part of what you are receiving.

As you move the control away from the center position, you will hear the high

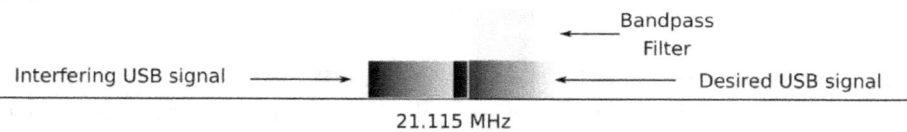

Figure 10.5: Narrower Band-Pass Filtering

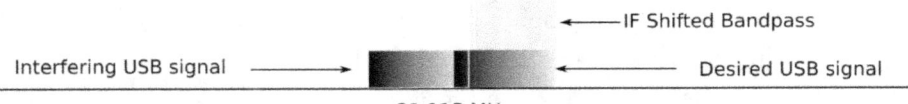

Figure 10.6: IF-Shift

frequencies disappear, leaving only the lower ones. As you move it in the other direction, the lows will disappear, leaving only the high frequencies. It sounds like you're adjusting a "tone" control, but you're not just altering the audio, you're altering what's being processed through the receiver in an attempt to avoid interference from stations very close to the receive frequency.

I) The Mode switch lets you select CW, Lower Sideband, or Upper Sideband. There's also a selection called "Tune" which switches the radio to very low power for purposes of adjusting those knobs that control the vacuum tube circuits.

J) Mic Gain (microphone gain) is where you adjust the loudness of your transmitted audio. You set it by keeping an eye on that meter up above while you transmit. Some radios show you your audio level – technically, your modulation level – on that meter, but this one does not. Instead, you set the meter to "ALC", the Automatic Level Control, and make sure it is "pumping" just a little but not too much. Starting out, just set this control in the middle; people will let you know if your audio is too soft (likely) or too loud (less likely.)

K) "NAR" stands for "narrow." That switch controls a band-pass filter. On this radio, those filters were optional, so that button might be hooked to a very narrow CW filter, a 2.5 kHz (or so) wide SSB filter, or no filter at all. The next switch down, PROC, switches the audio processing – also known as ALC -- for the outgoing audio on and off. The bottom switch, RF ATT, is RF Attenuation. It switches in a circuit that reduces the level of the incoming RF. It's useful if a signal is overloading your receiver.

L) Every HF transceiver I've ever seen has been equipped with some sort of multi-purpose meter to let you know things are working as they should. I can't imagine transmitting without having an eye on a meter, and you'll probably get to be the same way.

Not all transceiver meters read the same values. On this one, when the transceiver is in receive mode, the meter automatically reads received signal strength. In transmit mode, the meter is controlled by the switch below it and can read the output power, the final amplifier's plate current, the ALC level, or the high voltage power supply voltage. All-transistor transceivers will have different values. My

IC-746, for instance, has bar graph style displays on the LCD screen for received signal strength, power percentage, ALC and SWR.

M) Up there in the upper left-hand corner are the on/off switches and adjustments for two unrelated functions. One is the noise blanker, the other is VOX.

A noise blanker helps reduce pulse-type noise, such as ignition noise. It doesn't have any effect on continuous noise such as hum or interference like that generated by some LED and CFL light bulbs. The sensitivity control under the on/off switch adjusts what level of noise pulse will trigger the noise blanker.

The VOX control is used for hands-free operation. It stands for Voice Operating Transmit. It listens for you talking (or any other sound the microphone can hear) and when it hears something, it switches the radio into transmit mode. Use at your own risk! Just below the on/off switch for VOX is a sensitivity adjustment for VOX. That adjusts how loud or soft a sound will switch on the transmitter.

N) Through the middle of the left side of this radio are five controls. From left to right, the first is a switch called the Standby Switch. If it is set to REC, for Receive, the radio acts as you'd normally expect; it is in receive mode until you press the PTT button, then it switches to transmit mode. In SEND, the radio is continuously in transmit mode. In the days of vacuum tube transceivers that had to be hooked up to a dummy load and retuned with each frequency change, this switch got quite a workout. These days, with all-transistor rigs, they're not used as much.

Next is VOX Delay. This adjusts how long a silence the VOX circuit can hear before it drops the transceiver out of transmit mode. If you do decide to use VOX, for goodness sakes, adjust its delay properly so you won't be one of those hams who sounds like, "Well, ahhhhhhhhhhhh, I'm going to go out and, ahhhhhhhh, start the barbecue now, ahhhh…" That's the classic sign of someone "keeping the VOX open." Instead, just adjust that delay control so you can take a breath without switching out of transmit mode.

To the right of VOX Delay is a knob marked CAR. This one is not found on all radios, but if you find one, it's used to set the carrier level in CW mode. In essence, it's an output power control for CW.

The switch that controls the input to the meter above is next, and the switch that controls the AGC – Automatic Gain Control – is to the right of it. AGC is to received signals what ALC is to transmitted signals. In other words, it is supposed to level out the volume of received signals. This one can be set to off, to fast action, for CW signals, or to slow action, for SSB signals. AGC's make tuning across a band listening for signals a lot more comfortable. Without it, you either have to keep the volume down so low that you'll miss faint signals or risk blasting your ears when you come across a strong signal.

O) Finally, there's the RF Gain control. That controls the amount of amplification the received signal gets in the RF amplifier section of the receiver. My guess is most hams with average setups like mine seldom touch that control. We're not burdened by excessive received signal strength any more than we are burdened by excessive health or money. We just want the maximum amount of received signal we can get.

There really are just a handful of things we can adjust on a transceiver.

- The output power of the transmitter.
- The transmit and/or receive frequency.
- The transmitted bandwidth.
- The width and frequency range of the receiver's passband.
- The operating mode (CW, SSB, etc.)
- Noise reduction, such as the Noise Blanker.
- The speaker/headphone volume.
- The volume and audio processing of the received signal.
- The volume and audio processing of the transmitted signal.

There are sometimes multiple ways of accomplishing those adjustments, but those are really all the things we can possibly adjust.

Chapter 11

Operating on HF

Okay – so you have your antenna, your feed line, your radio, probably a power supply, and maybe a tuner. You've read the manual for your radio, even though it was about the size of *War and Peace*. You've read and followed the chapter on grounding and bonding your station, and you've hooked everything together. You, somehow, got an antenna analyzer connected to your antenna system and everything checked out fine so your transceiver won't melt 20 seconds into your first transmission. Excellent! By the way, through that whole process, you weren't "getting ready" to do ham radio; you were *already doing* ham radio!

Now for the actual getting on the air part. Let's play radio! Uh oh …. What are we gonna do?

Very broadly speaking, there are three families of modes widely used on the HF bands – CW, SSB phone, and digital modes. I don't mean to short-change any of the other modes, such as AM or slow-scan TV, but those are more specialized than what we'll cover here.

CW

We'll fly over the CW (Morse code operation) mode pretty quickly– I suspect if someone is motivated enough about HF to learn to use Morse code competently, that person is unlikely to be reading this book. If you're interested in learning CW, by all means download one of the smartphone apps such as Morse Machine and start practicing.

CW is particularly useful in these days of difficult propagation conditions – almost nothing punches through like CW. You also find a different crowd of people in the CW world. There are a couple of spots on the phone bands – especially 20 and 40 meters – that are crowded with a lot of --- well, let's put it delicately, "less than civil" conversation. You won't run across that stuff if you're working CW – but you also can easily avoid it on SSB by twisting the big knob and moving on to another spot on the dial.

SSB Phone

SSB (Single Sideband) phone is probably the most popular mode on HF, and there are lots of different ways to use SSB phone – contests, nets, DX'ing, and "search-and-pounce" – where you tune around the band until you find another human, then start a contact with them.

Let me suggest that before you pick up that microphone, you spend some time just listening to the HF bands. If you have internet access, you don't even need a radio to do that. Navigate to WebSDR

http://websdr.org

You'll be able to access Software Defined Radios from all over the world. As I write this, I'm listening to 14.236 MHz, on the 20-meter band, via a software defined radio that is located in Utah. It's a good propagation day for 20-meters, and a Slovenian ham named Slavko, S57DX, is making one contact after another with North American hams, mostly on the East Coast. Here's a typical exchange.

S57DX: CQ calling CQ Sierra five seven delta X-ray, sugar five seven Delta X-ray [some overseas DX'ers substitute "sugar" for "Sierra" – Slavko alternates between them.] calling CQ …

T32AZ: Tango thirty-two alpha Zulu. [T32AZ is a ham in East Kiribati, a tiny island nation in the middle of the Pacific, SSW of Hawaii.]

S57DX: Tango three two alpha Zulu, hello Ken. [Slavko probably has a computer screen open to QRZ.com or his log to quickly look up call signs.] You are 5 and 7, 57. ["Your signal is very readable and has a signal strength of 7.] My name is Slavko, Sierra Lima alpha victor kilo Ontario, over. [Yes, "Oscar" would have been the official NATO phonetic alphabet instead of Ontario. Slavko's English is somewhat accented, and he has probably learned that Ontario communicates better than his pronunciation of Oscar.]

T32AZ: Okay, Slavko, you are also 5 and 7, and thank you for the contact for the WAE this weekend. [WAE, "Worked All Europe," is the name of a radio contest.]

S57DX: Roger, Roger, Ken, I was very surprised, I was very surprised. In WAE contest I was with 100 watts, 100 watts, which … Right now I'm using full legal power, 1.5 kilowatts, and antenna is a five over four stack [five element Yagi sitting on top of a four element Yagi] aiming to America right now. Good signal, thank you very much, Ken.

T32AZ: *Okay, have a good day.*

S57DX: *Bye, bye. Thank you. QRZ?* ["Who is calling me?" In other words, "was someone trying to reach me during that last conversation, or does someone else want to contact me now?"] *Sierra five seven delta x-ray.* [And he's on to the next contact.]

At the moment, Slavko is chasing his WPX (Worked All Prefixes) award, which one wins by making contact with a certain number of different call sign prefixes. For instance, if he made contact with me, and I confirmed the contact, he'd have scored an AF7, but he might still be looking for AF1, AF2, etc. That's probably why there's no location information being exchanged. Slavko seemed very genial over the air, but he's not wasting a lot of time on idle chit-chat! He wants to complete the exchange of call sign, first name, and signal report in a timely fashion. Those elements plus your location will be a part of almost every conversation you have on ham radio, whether on VHF/UHF or HF, and regardless of whether the contact turns into a longer conversation or not. When you first start listening to HF, you won't catch everything that's going on, but soon you'll realize that most contacts contain those same elements.

I'm going to assume you know your call sign, first name, and location; maybe you have even looked up your maidenhead grid square on qrz.com. However, you might be a little puzzled about that signal report business.

On HF, hams use the "RST" system, shown in Table 11.1, page 175. If you read the table, you'll see the standards are very subjective. What this means for you is *there are no wrong answers* for this! When you give a signal report, you are giving your best estimate; that's all.

For phone, a signal report consists of two numbers. The first is for the "readability" of the received transmission; in plain language, could you hear what they were saying or not? That one you just judge by ear. The second number is for "signal strength", and, at least in theory, is what the Signal Strength meter on your transceiver says is the strength of their signal. That could range from 1 to 9, or even go as high as "9 plus 60." If someone was perfectly readable with an extremely strong signal, a typical signal report would be, "You are 5, 9 in Lake Stevens, WA." Someone you can only understand with considerable difficulty, with only a moderately strong signal might be "3, 7". If you get totally flustered and can't remember any of these fancy RST scales, "4, 7" is always a safe choice! (Not that either of us has ever done anything like that. So far as you know.)

You should also know that hard-core DX'ers and contesters very seldom give deeply thought-out signal reports. For them, everybody is "5, 9"; they're trying to get to that next contact, and everyone playing their game knows the signal reports are bogus. On the other hand, if you're in a conversation with a ham who has just, say, installed a new antenna, you'll want to take more care with your signal report.

HF Nets

Once you've spent some time listening, it's time to switch on the transceiver, take a deep breath, and start speaking on the microphone. I'd suggest a good starting point for HF operating is the same as for VHF/UHF; nets. There are lots and lots of them, on all the bands. I did a quick search the other day for 75-meter nets on theARRL's Net Directory Search and turned up 144 nets. When I searched all bands I got 420. Some are specialized – there's the Medical Reserve Corps net that only fires up on an as needed basis – and others are "y'all come" on very regular schedules – some even daily. Those 144 nets, by the way, are listed as "wide coverage" nets – when you look regional or local, or on more bands, you'll find even more choices.

The reason I suggest nets as a good starting place is because you can look up exactly when they'll be on the air and on what frequency. That saves you a lot of fruitless hunting. Pick a likely looking suspect for you and tune in and listen. Listen from the start, because that's where they'll explain their procedures.

I know the license exams leave the impression with some people that HF operation is wildly different from VHF/UHF operations. There are some differences, and a little vocabulary to learn, but let me set your mind at ease; the little eccentricities of HF communications all came about because they made communication work better. You're going to find that at least 90% of it is plain English; on nets, that number is probably closer to 99%.

There aren't many Q codes used on phone. All those Q codes and prosigns that were on the exams are mostly for CW users. The most commonly heard on phone are listed in Table 11.2. Usually, on phone, Q codes are used informally, almost as a form of slang. Rather than saying, "Man, the static is awful tonight", a ham might say, "I have tons of QRM tonight." That Q code didn't save any time, nor did it enhance communication. No one will care if you just say you have a lot of static instead of saying QRM. Just for confidence, though, you might want to keep a more complete cheat sheet of the Q codes close at hand. One source is the *Fast Track Ham Radio Facts* book; it has all the HF jargon and a bunch of other stuff in it.

Pay attention to the flow of the net, and you'll be up to speed in very little time.

The RST Signal Report System	
Readability	
1	Unreadable. "I can tell there's a signal there, but I have no idea what you are sending."
2	Barely readable. Occasional words are getting through.
3	Readable with considerable difficulty.
4	Readable with almost no difficulty.
5	Perfectly readable. "Sounds like you're right next door!"
Signal Strength	
1	Faint signal; barely perceptible.
2	Very weak signal.
3	Weak signal.
4	Fair signal.
5	Fairly good signal.
6	Good signal.
7	Moderately strong signal.
8	Strong signal.
9	Extremely strong signal.
Tone (Applies Only to CW)	
1	60 Hz AC or less, very rough and broad.
2	Very rough AC; very harsh and broad.
3	Rough AC tone, rectified but not filtered.
4	Rough note, some trace of filtering.
5	Filtered, rectified AC but strongly ripple modulated.
6	Filtered tone, definite trace of ripple modulation.
7	Near pure tone, trace of ripple modulation.
8	Near perfect tone, slight trace of modulation.
9	Perfect tone, no trace of ripple or modulation of any kind.
Additional Descriptors for CW	
X	Tone sounds very steady, like crystal (XTAL) control.
C	Tone is chirpy.
K	Tone is clicky.

Table 11.1: RST Signal Reporting System

Common Phone Q Codes		
Sign	As a Question	As an Answer or Notice
QRL	Are you busy? Is this frequency being used?	I am busy.
QRM	Do you have interference?	I have interference. Informally, "interference"
QRN	Are you trouble by static?	I am troubled by static. Informally, "static."
QRO	Shall I increase power?	Increase power. Informally "I am operating with high power."
QRP	Shall I reduce power?	Decrease power. Informally, "I am operating with low power."
QRZ	Who is calling me? Informally, "Who wants to talk with me next?"	You are being called by [call sign]
QSL	Do you acknowledge receipt (of the message)? Informally, "do you understand?"	I acknowledge receipt, or "I understand."
QSO	Can you communicate with [call sign]?	Formally: I can communicate with [call sign]. Informally, a conversation on ham radio.
QSY	Shall I change frequency?	I am changing frequency to [frequency].
QTH	What is your location?	My location is [location].

Table 11.2: Common Phone Q Codes

Formal nets provide a structured environment that can be great for a beginner – you know when to talk, you know, basically, what sort of communication is expected, and you have a sense of the rhythm of the net.

HF Contesting

I really don't suggest starting right into contesting, unless you have a good Elmer or Elma by your side. You'll get your skills up to speed quickly enough and be able to play that game, but – well, some of the players take the game really, really seriously, and if you're not right on top of the contest exchanges, they get all in a twist. Let's face it, some people kind of lose their minds in the heat of a competition. Heaven help you if you slow down their pace, or worse, fail to log your QSO!

The good news about contesting on HF is that, unlike VHF/UHF, there are tons of HF contests. For example, the current issue of QST, the ARRL monthly magazine, lists 61 contests occurring this month. Nine of those 61 occur at least partially on VHF and/or UHF frequencies. The rest are for HF only.

We covered strategies for contests in the VHF/UHF section of this book, page 128, and all those strategies apply to HF as well. I'll just emphasize again, it's all about speedy, efficient operation.

Search-and-pounce on HF

Search-and-pounce is great if you have a little more patience and don't care so much about the structured environment of a net. Your best bet is to try to tag on to the end of a conversation. That way you'll have the call signs straight – and written down. It's easy enough to tag on – wait till they're wrapped up, then say one of their call signs, and your call sign. Remember to hit those phonetics – the real ones! They need a moment to get your call sign straight, too.

You'll be able to tell from the conversation that just ended what type of operation that operator is into. If they've had a long, casual conversation covering several topics, they're ragchewers and looking for more of the same. If it was "call sign, location, signal report, OVER", they're probably contact collectors of one sort or another, working on their Worked All States, or something similar, so all they want is the same from you.

DX on HF

DX, the prosign for "distance", involves contacts with amateurs in other countries or "DX entities." Alaska, for instance, counts as DX from the continental US, as does Hawaii.

The truth is, your first DX contacts will probably be more or less accidental. You'll be chatting with someone in the US and, boom, the band will open up and suddenly you'll be chatting with someone in Argentina. Log that contact; you only have 99 more DX regions to go to get your DX Century Club award!

Speaking of logging, you'll probably want to get yourself set up on Logbook of the World. LotW, as it's known, is a service run by the ARRL as something of an electronic QSL bureau. It does things like tracking your confirmed QSL's toward

UNITED STATES
AF7KB

Michael Burnette

Lake Stevens, WA

CONFIRMING QSO WITH KC7YL					
RADIO	DATE	UTC	MHz	MODE	RST
IC-746	08/31/2019	1842	14.313	FT8	N/A

Pse QSL Tnx   73,

Figure 11.1: QSL Card

various awards, such as Worked All States or that DXCC. There's a confirmation process to complete and you'll need to install a program known as Trusted QSL (TQSL for short) which assures that only you can make entries on your log. To get started with LotW, which you can do any time after you get your first amateur radio license, navigate to this address:

https://lotw.arrl.org/lotw-help/getting-started/

There's a little bit of a learning curve with LotW, but it's not nearly as complex as that page makes it look, I promise.

Of course, you're more than welcome to confirm your QSL's the old fashioned way, with QSL cards sent through the postal service. That has grown less popular through the years, but it is the *only* way some hams out there will confirm a contact. Figure 11.1 is a sample QSL card I made in under a minute at

http://www.radioqth.net/qslcards

Your own QSL card can be as elaborate or simple as you like, but you'll want to include at least the date, time (in UTC), frequency, and mode of the contact. I've RST (Readability/Signal Strength/Tone signal report) as n/a for this FT8 contact, but in other modes, it's good to include that signal report as well.

If you start to get more serious about making DX contacts, it is useful to learn as much as you can about ionospheric propagation and how to take advantage of it.

In general, you'll find that the shorter wavelength bands perform better during the day while the longer wavelength bands perform better at night. When I say "during the day" and "at night," I mean day or night *between you and the other station*. For instance, early morning here on the West Coast might be a good time to talk with Japan on 75 meters, but a time to talk with Pennsylvania on 20 meters.

One software resource for predicting propagation more precisely than that is known as DXLab Suite from David Bernstein, AA6YQ. For DX'ers, DXLab Suite is extraordinarily useful, and is free! The program is for Windows only. Download at

http://www.dxlabsuite.com/dxlabwiki/InstallLauncher

DXLab Suite has many useful capabilities. It's a suite of programs whose uses range from propagation prediction to operating your computer-controllable transceiver. We won't even begin to cover all the ways of using DXLab Suite here.

Because it has lots of functions, there's certainly a bit of a learning curve to using DXLab Suite, but there's an active online user community and tutorials. Here's one way to use it.

Let's say I want to QSO with a ham in Ireland, EI2KC, whom I happen to have seen in a YouTube video.

I open the parts of DXLab Suite called PropView (Propagation View) and DXView. Figure 11.2 shows the window of DXView.

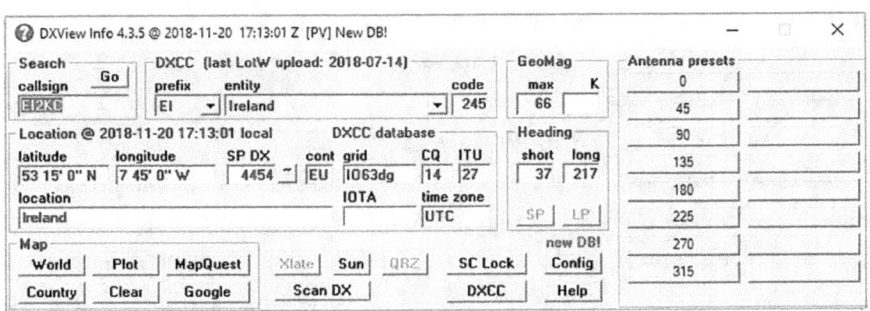

Figure 11.2: DX Lab SuiteDXView

The only entry I made is "EI2KC" in the call sign. The program filled in everything else, including his latitude, longitude, grid square, and even the headings for the short path and long path to him. It did it almost instantaneously, too.

Don't have a particular call sign in mind? Not a problem! In the lower left corner you see the "Map" section of DXView. Click on "World" and a map like the one in Figure 11.3 opens.

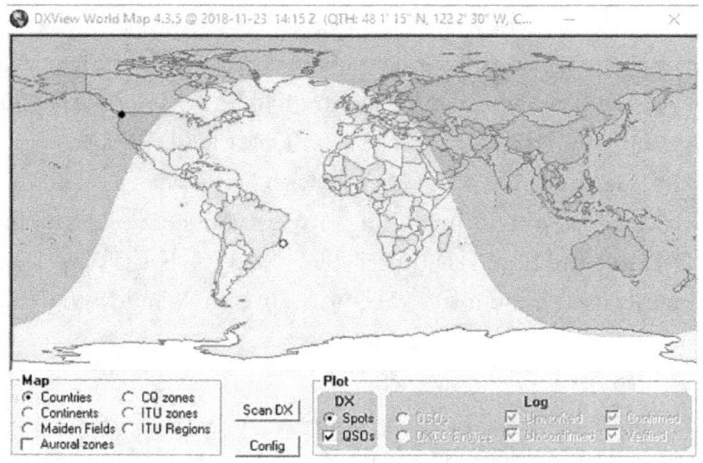

Figure 11.3: DXView World Map

Click on a destination, and those coordinates auto-fill in the DXView window.

The coordinates in DXView transfer automatically to PropView, Figure 11.4 which is the part of the program that's going to create your propagation forecast.

The PropView window looks up the latest space weather numbers with the click of a button and puts them in the "Conditions" section: the Solar Flux Index, the Sunspot Number (in this case the *Smoothed Sunspot Number*), and the K index. You pick the "Avail" number; that's a measure of, basically, how picky are you about the probability of the path being measured actually working. You also pick the desired SNR, the Signal to Noise Ratio, and/or the mode you want to use. SSB demands less noise than CW, for instance.

There are also choices you can make about the direction of the prediction – from you to them or vice versa; the noise level at your QTH, the noise level at the target QTH; your power level and theirs; and, long path or short path.

Then you click the "Predict" button and the program generates a propagation prediction for the path between you and the target you have chosen. It looks like Figure 11.5.

The graph covers 24 hours. The X axis is time-of-day in UTC, with the vertical line showing the current time. The bars beneath the time scale indicate day and night at your QTH and at the target.

The vertical axis is frequency. The left-hand vertical axis is calibrated in MHz, the right in wavelength.

The line graphs indicate the predicted critical frequency – not to be confused

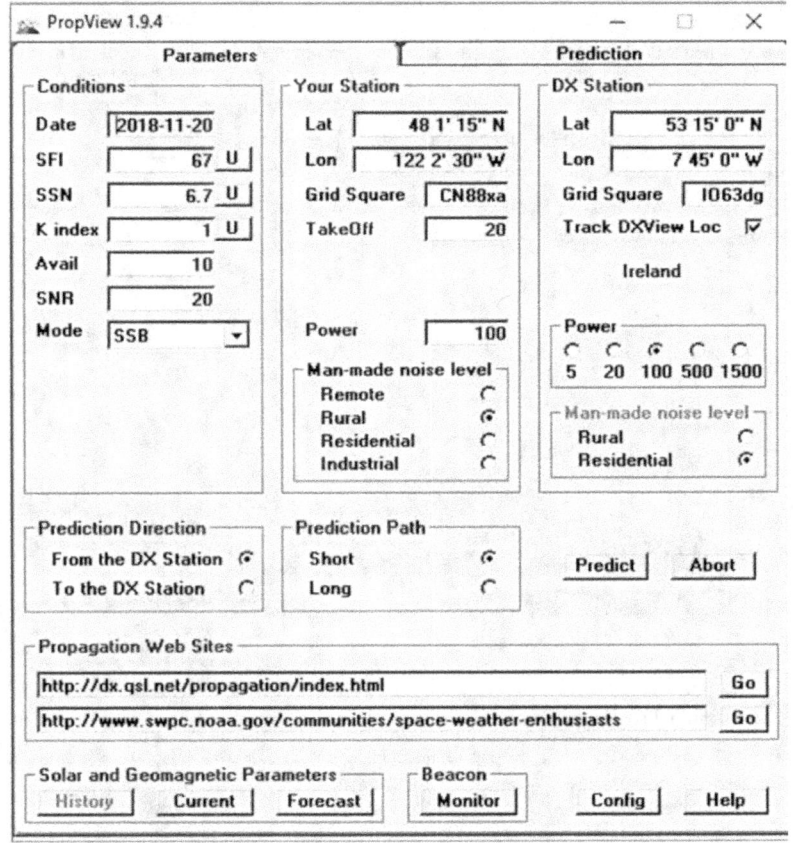

Figure 11.4: DXLab Suite PropView Window

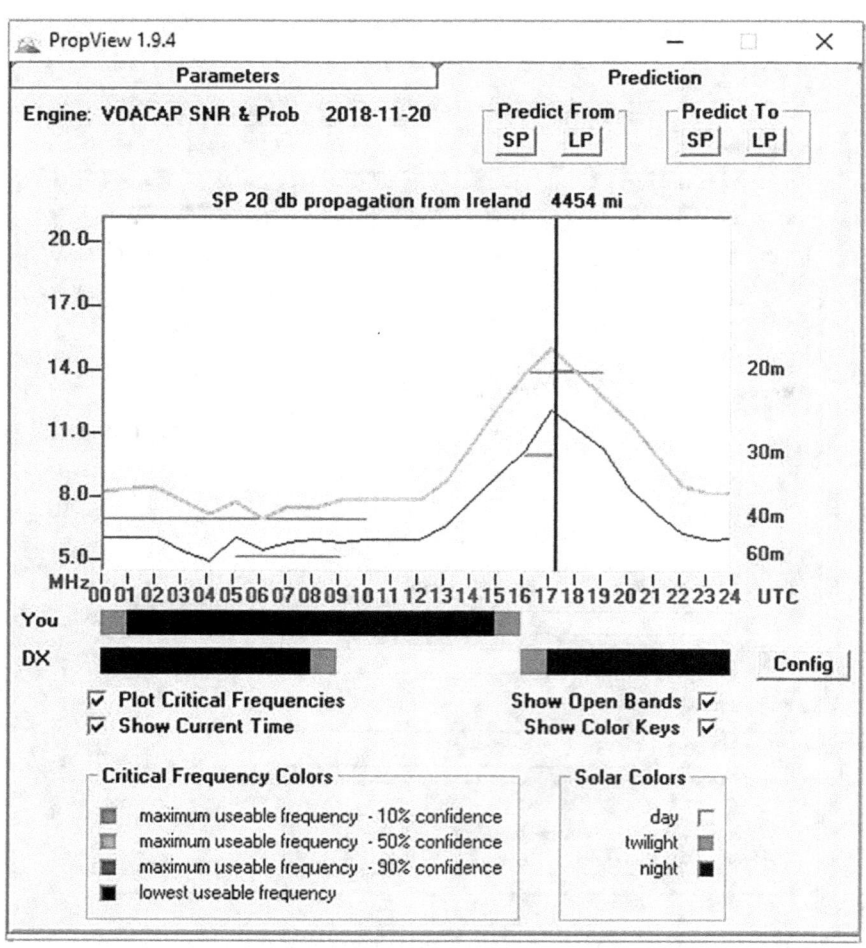

Figure 11.5: PropView Propagation Forecast

with the MUF. The top line predicts the critical frequency with 50% confidence, the bottom line predicts it with 90% confidence. The program also generates a critical frequency prediction with 10% confidence, but because I have "Show Open Bands" checked, it isn't displaying that one, since it's up in some closed band.

Sometimes the chart will show you the LUF as well, but in this case it was below the ham bands.

The horizontal lines on the chart are the ones you want to see a lot of. They indicate band openings, and they get wider as your chances on the band get better. You can see that at the moment I made this prediction, I had about two hours of 20-meter opening left before the band was predicted to close.

(It might be useful for you to know that the actual graphs produced by the program are color-keyed – if you're holding the print edition of this book, you're obviously not seeing all the colors, which make these predictions far easier to interpret.)

Remember, though, radio waves are always propagating *somewhere* and even the most sophisticated software can't tell you what your results will be today until you switch on the radio and start trying.

Digital Modes on HF

Another toy in our toy box is the digital modes. There are lots of them, many designed for specialized purposes. I'll touch on two of the most popular. One is more of a chat mode, a lot like RTTY or AOL Instant Messenger. A more social mode, if you will. The other is wildly efficient and strictly about "Call sign, signal report, 73" and on to the next contact.

First, let's talk about setting up for digital. You'll need a computer running a reasonably modern operating system. You don't need any sort of high-end gaming computer, though. My shack computer is an aging laptop and believe me, its CPU is made of wood. It's running Windows 7. The computer has to get audio to and from the transceiver.

Figure 11.6 shows what I'd call the "textbook" set-up to accomplish that.

The headphone output of the computer is running into the microphone input of the transceiver. The headphone output of the transceiver is running into the microphone input of the computer.

There are some inherent problems with this setup. First, there's no "Push to Talk" control anywhere in that loop, so you either have to manually switch to transmit mode or do what we do which is run with VOX switched on.

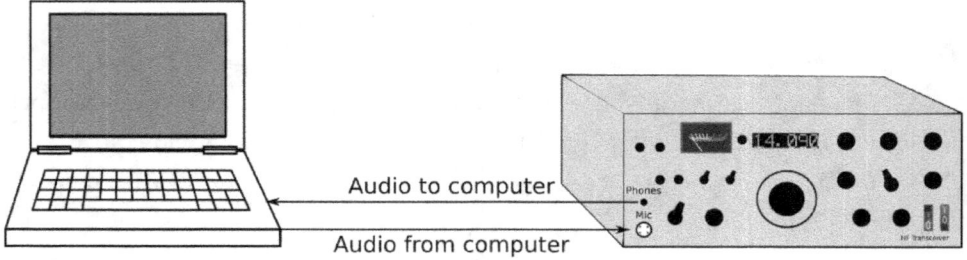

Figure 11.6: "Textbook" Setup for HF Digital

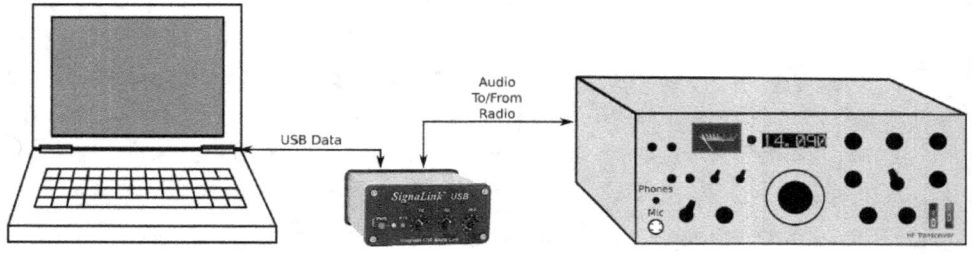

Figure 11.7: Real World Setup for HF Digital

Second, the audio outputs of both devices are at line level – around one volt – and the microphone inputs are expecting a signal at mic level – typically around a millivolt. That's 30 dB of difference, so setting volumes properly is often a delicate operation.

I guess there are some people running with that textbook setup, but I don't know any of them. Figure 11.7 is what I'm running, and more or less what everyone I know is running.

That box in the middle is called a SignalLink. It replaces your laptop's sound card. It plugs into a USB port on the computer, and into your radio – somewhere. There is no such thing as a "standard" plug for connecting external gear to the transceiver, so it could be anything from a ¼" phone jack to a USB port. The maker of SignalLink, Tigertronics, makes a wide assortment of cables for the connection to your transceiver.

Now that your computer and radio can communicate with each other, the next step is to get the right software running. For the first mode we'll talk about, PSK31, fldigi is free software for that mode, and there are other programs that will also do PSK31. You can download the latest version of fldigi at:

http://www.w1hkj.com/

Ham Radio Deluxe, mentioned in the section on remote station operation, page 155, also does PSK31 – if you're not familiar with Ham Radio Deluxe, it's quite the software package. It does your logging, handles multiple digital modes, integrates with the WSJT software suites, and a whole lot more.

https://www.hamradiodeluxe.com/

Ham Radio Deluxe is not a free program – at the moment, it's $99.95. (WSJT stands for "Weak Signal Joe Taylor". It's a suite of programs to transmit and receive various digital formats, all designed for weak signal conditions.)

For the very weak-signal modes that are very popular, at least at the moment, including FT8, it's the WSJT-X suite of programs, another free program.

Let's start with PSK31, the one that's a lot like RTTY, radio teletype. It's a keyboard-to-keyboard conversation mode. PSK31 means Phase Shift Keying, 31 baud, so that describes the modulation mode and the data speed.

Let's say propagation is looking at least fair on the 15-meter band. You'll tune your radio to 21.230 MHz – that's the unofficial PSK frequency for that band, and you can find all the frequencies for the different bands easily on the web with Google. Set your radio for USB – unlike RTTY, all the digital modes run on USB. Some won't work any other way, but the rest run there by convention. You probably will hear a lot of static on the frequency, but don't worry about it; PSK31 is made for fair to poor propagation conditions. You fire up the PSK31 software of your choice – let's say you opted for Ham Radio Deluxe. You'll get a screen like the one in Figure 11.8. Fldigi's is similar, by the way, but with a lot fewer snazzy features.

PSK31 has a very narrow bandwidth – it's skinnier than CW. So, there are going to most likely be a lot of signals right around 21.230. Since it operates with audio tones on USB, different audio tones create different offsets from 21.230, and you can see about 8 signals in the 3 kHz of waterfall display there. Your computer will handle the offsets – you'll leave your radio set to 21.230. When you click on the waterfall display, your software will start decoding that conversation, which you see in the window above.

You can sit and monitor PSK31 conversations until you get the hang of how they go. While the software can be a little confusing at first – at least it was for me, but I'm easily confused – the operating procedures are quite straightforward. When you get your nerve up and you find someone calling CQ, answer them! You're off and running on PSK31.

PSK31 conversations use a lot of lower-case letters and abbreviations. The lower case is because PSK31 uses what they call "varicode" – you might remember it from your license exam. Like Morse code, it uses very few symbols to represent

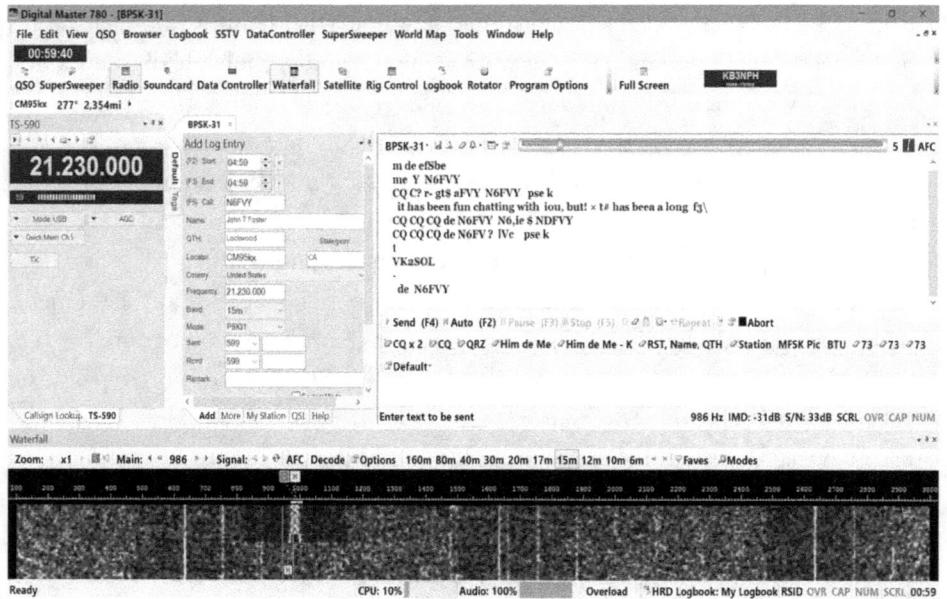

Figure 11.8: Ham Radio Deluxe PSK31 Screen

the most common characters, so lots of capital letters slow things down. You don't see it in this picture, but there's a lot of the same sort of abbreviations as you'd find in CW or RTTY. I think any of us who text on our phones much are used to sentences like "OK LOL C U LTR" – you'll catch on quickly enough, since the same ones get used over and over.

As you can see from the picture, one doesn't have to be an expert typist to use PSK31 – in fact, there are macros built into most of the software packages that automatically send things like CQ CQ CQ de AF7KB and other common text strings. PSK31 chugs along at a send rate that's about equal to average typing speed – you don't have to be super-typist at all.

You can see a great video by Dave Casler, KE0OG, of a sample QSO on PSK31 if you look up PSK31 on YouTube.

The bad news about PSK31 is that so many people have migrated over to using FT8 that PSK31 conversations are getting a little scarce – but they're still out there, and I think they'll probably pick up as propagation conditions improve.

Then we come to the latest officially released weak-signal mode, the notorious FT8.

As a technical achievement, FT8 is amazing. If PSK31 is a weak-signal mode, FT8 is an almost-invisible signal mode --- it's quite capable of pulling out signal with a signal-to-noise ratio in the single digits. It's specifically designed for poor propagation conditions, using low power through a marginal antenna. You do not need to fire up a big ol' 5 element Yagi on a 150-foot tower with 1500 watts to make this stuff work worldwide. The first time I ever tried it, I was running about 50 watts on 20 meters into a Buddipole set up on our patio – the mast was only about 12 feet up because it was a windy day. Now, Buddipoles are great, and we've had a lot of fun with ours, but that is not what you'd call a Big Dog setup for 20 meters. It was a rotten propagation day – Paul Hermann's solar weather widget said POOR conditions on every band at all hours, and believe me, it was right. I fired up the computer and tuned in 14.074. I flailed around with the software for a little bit, got it working, and had a contact within a couple of minutes. It was – okay, it was all the way to Marysville, Washington, about five miles from my house, but, hey! Contact! After that I had Wisconsin, Michigan, Belize, and Portugal, all within about an hour.

FT8 comes to us from the great guru of weak signal modes, Joe Taylor, K1JT and Steven Franke, K9AN. In fact, FT8 stands for Franke-Taylor, 8 tone frequency shift keying

Your radio/computer setup for FT8 is the same as for PSK31. You'll be running that WSJT-X weak signal suite, which you can download from Joe Taylor's site at Princeton University. Figure 11.9 shows a screenshot.

On the left you see all the activity in the band – and when it's live, you'll see incoming calls come in the bottom. If you squint, you can see WA6GXQ is calling CQ – it's here, second from the bottom. If I double click on that, that call sign will show up over in the right hand window, and my computer will send a response to that CQ. This operator has the software set to Auto Sequence, so if WA6GXQ responds to my answer, my computer will just step through the standard messages in this window here – it even auto-generates a signal report, and wishes the other operator a pleasant 73 when we've finished exchanging signal reports. That's about it for in-depth, intimate conversation on FT8! FT8 is about making contacts – period. It's possible to hack in tiny little messages, but I've never personally seen it done, and doing so sort of defeats the efficiency of the mode, anyway.

One thing you must do for FT8 is synchronize your computer's internal clock to a good time standard. There's a web site called time.is (screenshot in Figure 11.10) that will help you with that, and there are programs you can install that will automatically update your clock.

Windows 10 does a pretty good job of keeping your time accurate but do use time.is to check it out before you start transmitting. The reason for the accurate

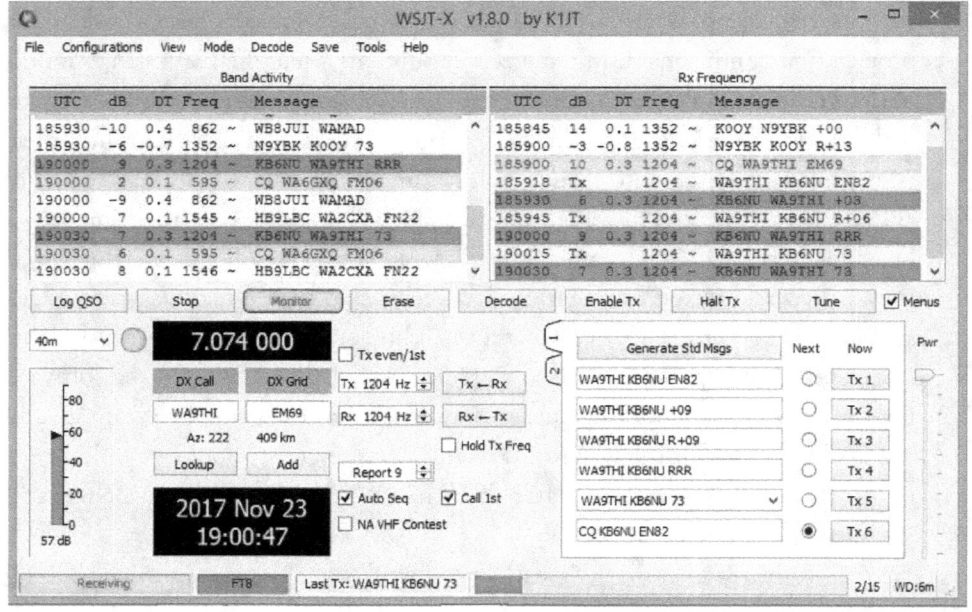

Figure 11.9: WSJT-X FT8 Screen

time is that FT8 operates with precisely timed 15 second slots of listening, then transmitting, then listening again, etc. If you're not synched up, you'll be transmitting when the other station isn't listening – and vice versa.

Figure 11.10: Time.is Screenshot

Chapter 12

Closing Thoughts

We hope we've provided enough encouragement and know-how for you to start taking action on your ham radio dreams. It's fun to know about ham radio things, but it's a lot more fun to put that knowledge into action, and we've tried to inspire you to do that.

When we said to consider us your Elma and Elmer, we were serious. If you're stuck, if there's something you need to know to move forward, drop me an e-mail! You'll find me at:

af7kb@fasttrackham.com

If you are a perfectionist, I'm afraid I have some bad news for you. I know you folks – you're very concerned that you're going to make errors and everyone will know you're a beginner. Well – brace yourself – even after completing this book, you still will make errors and everyone will still know you're a beginner. Happily, precisely 95.2% of all hams are good-humored and forgiving, especially if they know you are new to the hobby. As for the other 4.8% – well, you weren't going to make them happy, anyway, and they'll survive their traumatic encounter with the neophyte. Forget about perfection – this hobby's about fun!

Get out there, get on the radio, have some fun, and do some wonderful things. There's a whole world of community, camaraderie, and challenge waiting for you.

Amateur Radio Glossary

Term	Definition
\multicolumn{2}{c}{0 - 9}	
10 codes	Non-standardized prosigns -- over-the-air shorthand -- used by emergency responders and Citizens Band hobbyists. Not generally well received nor broadly comprehended in the amateur radio community.
72	73 in QRP operation.
73	Prosign for "best regards" or "best wishes."
88	Prosign for "love and kisses."
\multicolumn{2}{c}{A}	
AC (alternating current)	An electrical current which flows first one direction then the opposite and, typically, continues to do so.
Adaptive filter	A digital filter used in digital signal processing.
ADC	Analog to Digital Converter. In amateur radio, most commonly used to turn audio or RF into digital data.
Adjacent-channel interference	Interference received from a nearby frequency.
Admittance	The reciprocal of impedance. Measured in siemens.
AF	Audio Frequency.
AFC	Automatic Frequency Control. A circuit that automatically compensates for frequency drift.
AFSK	Audio Frequency Shift Keying. An analog modulation method for transferring digital data such as RTTY text.
AGC	Automatic Gain Control. A circuit that automatically optimizes receiver gain.
AGL	Above Ground Level, for antenna installations.
Airlink Express	Multiple digital HF mode software.
ALC	Automatic Level Control. A circuit that automatically limits RF drive to the final amplifier to prevent distortion.
AM	Amplitude modulation. While SSB is a form of amplitude modulation, in amateur radio use, AM is normally double sideband modulation with a full carrier.

Continued on next page ...

Term	Definition
Amateur operator	A person holding a written authorization to be the control operator of an amateur station.
Amateur Service	"A radiocommunication service for the purpose of self-training, intercommunication and technical investigations carried out by amateurs, that is, duly authorized persons interested in radio technique solely with a personal aim and without pecuniary interest." - FCC Part 97
Amateur station	A station licensed in the amateur service, including necessary equipment, used for amateur communication.
Ammeter	A test instrument that measures the amperage passing through a circuit.
Amperage	The amount of current flowing through a conductor or circuit.
Ampere (A)	The unit of current flow. 1 ampere = 1 coulomb of charge passing a point in 1 second.
Amplitude Modulation	A system of transmitting information by modulating the amplitude of an RF carrier wave.
AMSAT	Amateur radio SATellite. The name for amateur radio satellite organizations world-wide, but in particular the Radio Amateur Satellite Corporation.
AMTOR	Amateur Teleprinting Over Radio. A form of RTTY.
Analog (communication)	The transmission of signals or information represented via a continuously variable dimension, such as amplitude, frequency, or phase.
ANF	Automatic Notch Filter.
Antenna	A wire, rod, pole, or other device for radiating or receiving electromagnetic waves.
Antenna analyzer	An instrument used to measure and display various characteristics of antennas at various frequencies.
Antenna gain	See "gain"
Antenna matching network	See "Antenna tuner"
Antenna party	A long-standing tradition among hams where several gather to assist a fellow ham in mounting antennas and/or towers.
Antenna switch	A specialized switch used to connect one transmitter, receiver, or transceiver to more than one antenna while connecting the other antenna(s) to ground for safety.
Antenna tuner	A circuit or device for matching the impedance of an antenna system to the impedance of the transmitter and/or receiver.
AOS	Acquisition of Signal from a satellite.
APC	Automatic Power Control. Limits current to final amplifier when SWR is high.
Apogee	The point in an orbiting object's orbit where it is farthest from the object around which is it orbiting.
APRS	Automatic Packet Reporting System. APRS combined with a GPS can provide real-time updates of a mobile or handheld transmitter's location.

Continued on next page ...

Term	Definition
ARC	Amateur Radio Club. For instance, the Mount Diablo Amateur Radio Club is known to its members and friends as MDARC -- "EMM-dark."
ARES	The Amateur Radio Emergency Service.
ARISS	Amateur Radio on the International Space Station.
ARRL	The American Radio Relay League. The national amateur radio club and lobbying organization - our voice at the FCC and Congress.
ASCII	American national Standard Code for Information Interchange. A 7 bit digital code for the transmission of text.
ATV	Amateur Television.
Auroral propagation	Propagation from auroras.
Autopatch	A system for connecting a repeater station to the telephone system for the purpose of allowing amateur operators to place phone calls via their radios. (Seldom heard or even existent these days.)
AWG	American Wire Gauge. A system of specifying the diameter of wire.
Azimuth	In the context of directional antennas, a direction in the horizontal plane. Most often shown on a graph of polar coordinates, with 0 degrees corresponding to the far right side of the circular polar graph and the degree scale running counter-clockwise around the circle.
B	
Backscatter	Radio signals bounced back toward the transmitter (and elsewhere) from, among other possibilities, ionized patches in the ionosphere, from a rain front, or from aurora.
Balanced line	A feed line with neither conductor at ground potential, such as ladder line.
Balun	A device for connecting an unbalanced source (such as coaxial cable transmission line) to a balanced load (such as a dipole antenna.) A balun can also be a choke balun, used to prevent unwanted RF from traveling back down a transmission line into the shack.
Band	A range of frequencies.
Band spread	A receiver specification denoting how far apart on the dial stations on nearby frequencies will seem to be. Somewhat like an RF analog of "zoom." Usually expressed as "kHz per turn" of the tuning knob.
band-pass filter	A circuit or device that allows a desired band of frequencies to pass while blocking all others.
bands are dead, The	"Communicating on HF using the ionosphere is more difficult today than I had anticipated."
Bandstop filter	A circuit or device that stops an undesired band of frequencies from passing while passing all other frequencies.

Continued on next page ...

Term	Definition
Bandwidth	The difference, in Hz, between the lowest and highest frequencies occupied by a given signal, or generated, passed, or successfully processed by a particular device. Technically, the width of a frequency band outside of which the mean power is attenuated at least 26 dB below the mean power of the total emission, including allowances for transmitter drift or Doppler shift.
Baofeng	Amateur radio equipment manufacturer, specializing in inexpensive handhelds and mobile radios. Chinese for "thunder." Pronounced BAO (like "take a bow") fung.
Battery	A device that supplies power by converting chemical energy into electrical energy.
Baud rate	A measure of the rate at which digital data is transferred. Equal to the maximum number of "symbols" that can be sent per second. Not equivalent to bit rate, nor to the number of text characters that can be sent per second. (See "symbol" for more detail.)
Beacon station	An amateur station transmitting for the sole purpose of observation of propagation and reception or other related activities. See NCDXF.
Beam antenna	Any directional antenna.
Beverage antenna	A very long wire antenna, normally for receiving only, mounted near the ground and terminated in a resistor to ground.
BFO	Beat Frequency Oscillator. An oscillator which supplies a signal to a receiver's detector circuit. Necessary to produce an audio tone for CW signals and for SSB.
Bipolar transistor	A type of transistor consisting of a layer of P or N type material sandwiched between two layers of the opposite type of material.
Bird	Nickname for an amateur radio satellite. Also the brand name of a high-end professional grade directional wattmeter.
BNC	Bayonet Neill-Concelman. A type of antenna connector. Also used for some test instruments such as oscilloscopes.
Boat anchor	A large, weighty piece of vintage amateur radio equipment.
BPS	Bits Per Second. A measure of the rate at which digital data is transferred.
BPSK	Binary Phase Shift Keying. A digital data modulation method.
Brass pounder	A CW operator who prefers a straight, mechanical key vs. a "bug" or electronic keyer.
Broadcasting	Transmissions intended to be received by the general public, either direct or relayed.
Bug	A mechanical device for generating the dots and dashes of Morse code semi-automatically.
Bunny hunt	Radiosport of finding hidden transmitters. See also "foxhunting."
Busy lockout	A feature on some transceivers that prevents transmit on a frequency already in use.

Continued on next page...

Term	Definition
C	
C4FM	The digital data modulation system used by Yaesu's Fusion digital voice and data system.
Call sign	A unique identifier assigned to someone who has attained an amateur radio license.
Capacitance	The amount of electrostatic storage available in a capacitor. Measured in Farads.
Capacitance hat	A system of wires or other structure at the top of a vertical antenna that reduces its inductance and increases its bandwidth.
Capacitor	A component that stores energy in an electrostatic field.
Capture effect	An effect exhibited by FM receivers when a stronger signal "captures" the receiver, blanking another signal.
Card Checker	A volunteer who validates QSL's for those applying for various QSL-based awards.
Carrier	The radio frequency sine wave that is modulated with the information to be transmitted.
Carrier frequency	The center frequency of the band of frequencies in an information carrying signal being transmitted or received.
Cartesian coordinates	See "rectangular coordinates."
Cavity filter	A type of very narrow-band RF filter. Typically found in repeater installations as part of the "duplexer."
Center-fed	Applied to antennas with a feed point at the center of two halves of the antenna. The "+" output of the transmitter connects to one half, the "-" to the other half, as in a typical wire dipole.
Chassis ground	A common connection for all components in a device that connect to the negative side of the power supply.
CHIRP	Free software for programming a wide array of transceivers with a computer. From http://danplanet.com
Chirp	In CW, a slight shift in the transmitter frequency each time the key is closed, which creates a characteristic "chirping" sound at the receiver.
Choke	A inductor used to block alternating or pulsed direct current, sometimes while passing steady direct current. One might use a "choke" on the power supply to a mobile radio to allow 12 V DC to pass while blocking intermittent noise pulses from the ignition or alternator whine; or, one might install a choke at an antenna to block RF from traveling down the shield of a coaxial cable feed line.
Circulator	A device in which an RF signal entering any port is transmitted to the next port only. Common in repeater installations.
Clarifier	Alternate name for Receiver Incremental Tuning control.
Clipping	Waveform created when an amplifier is overdriven.
Closed repeater	A repeater that can only be used by those in possession of a special code.

Continued on next page ...

Term	Definition
CLOVER	A full-duplex Phase Shift Keying data mode.
CNDX	Conditions.
Coax	Coaxial cable. Transmission line with one conductor inside another, the two being separated by an insulating layer known as the dielectric.
Code key	A device for sending Morse code.
Color code	In the context of transmission and reception, the form of modulation used to transmit information. For instance, CW, SSB, Phase Shift Keying, etc.
Condenser	Vintage term for a capacitor.
Conductivity	The reciprocal of resistance. Measured in siemens.
Conductor	A material that will readily pass an electric current.
Contesting	Creating contacts with as many stations as possible over a specific amount of time, and often on specific bands.
Control operator	The party responsible for the transmissions of a station.
Control point	The point at which the control operator function is performed.
Controlled environment (RF exposure)	An area where an RF signal may cause radiation exposure to people who are aware of the radiation and can exercise some control over their exposure. There are different RF exposure standards for controlled environments vs. uncontrolled environments.
Conventional current	The flow of electrical charge from a positively charged location to a negatively charged location. Until the discovery of the electron, it was thought electricity flowed from positive to negative, so all schematic diagrams use this convention. It is perfectly valid to think of an electric current consisting of positive charges moving toward negative, negative charges moving toward positive, or both simultaneously.
Coronal hole	Sunspot related solar activity that may enhance 10-meter and VHF propagation while degrading HF propagation.
Counterpoise	Precisely, a ground plane radial or radials mounted above the surface, but used more generally for any ground plane.
Coverage area	The area in which a station can be received.
CQ	Prosign for "Calling anyone." Can be used in CW or phone transmissions. Primarily used on HF and VHF/UHF simplex -- not commonly used on repeaters.
CQ Magazine	One of the two major general interest amateur radio magazines, the other being QST Magazine.
Cross modulation	Interference caused by the interaction of two (or more) carriers.
Crossband repeat	A mode available on some dual-band VHF/UHF radios allowing the radio to receive on one band and simultaneously transmit on the other.
Crystal	A crystal of silicon machined to resonate at a specific frequency when an electric current is applied to it.
CTCSS	Continuous Tone Coded Squelch System. A system of sub-audible tones used to control most repeaters. When the right tone is added to your transmission on the proper frequency, the tone "opens up" the repeater.

Continued on next page ...

Term	Definition
Current	The flow of electricity through a conductor.
CW	Continuous Wave. Transmission of information, almost universally by Morse code, by switching the carrier off an on. Colloquially, in amateur radio use, any Morse code transmission, or the code itself.
CW filter	A circuit in a receiver that limits the width of the IF passband to improve selectivity in crowded band conditions.
D	
D region	The lowest layer of the ionosphere, present only during daylight hours.
DAC	Digital to Analog Converter. In amateur radio, most commonly used to turn data into audio.
dB	Decibels. Not a unit of measure, but rather a logarithmic expression of the ratio between two values. Can be suffixed with various letters to indicate the basis of comparison. For instance, dBm is "decibels of difference compared to a 1 mW signal."0
dBd	The gain of an antenna relative to that of a 1/2 wave dipole antenna.
dBi	The gain of an antenna relative to an isotropic antenna.
DC (direct current)	An electrical current which flows only one direction. It may vary in intensity or even be intermittent, but it still flows in only one direction.
DCS	Digital Coded Squelch. An alternative to CTCSS for repeater access control.
Detector	A circuit in a receiver that retrieves the information from the received signal.
Deviation	A measure of the maximum frequency changes on either side of the carrier center frequency for an FM signal.
DF	Direction finding. Locating a source of a radio transmission through the use of directional antennas and associated equipment.
DigiPan	HF digital mode software.
Digipeater	A packet radio station used to relay digital packet messages.
Digital (communications)	Computer-based amateur radio modes. Can be data modes, such as packet or APRS, or text based modes such as RTTY or the WSJT modes.
Digital Radio Mondiale.	Digital Radio Mondiale. A digital mode designed to compress signals into reduced bandwidth. Not to be confused with DMR, a digital voice mode.
Diode	A solid-state device that acts as a one-way valve for electricity, among other applications.
Diplexer	A device for frequency-domain multiplexing. A diplexer allows two transmitters to use the same antenna on different frequencies simultaneously.
Directional wattmeter	See "wattmeter"
Director element	A passive antenna element, typically located in front of the driven element.
Dish	A highly direction parabolic dish antenna.

Continued on next page ...

Term	Definition
DMR	Digital Mobile Radio. A digital voice mode.
Domino EX	A digital mode that uses a modified form of MFSK known as IFK, for Incremental Shift Keying.
Doppler shift	Frequency shift caused by relative motion between a transmitter and a receiver. Common in satellite communications.
Downconverter	A device to convert a signal from a higher frequency to a lower one.
Downlink	Frequency used by a satellite to transmit to the user.
Driven element	The element in an antenna that is connected to the output of the transmitter.
DRM	Digital Radio Mondiale. A digital mode designed to compress signals into reduced bandwidth. Not to be confused with DMR, a digital voice mode.
DSP	Digital signal processing.
D-Star	Digital Smart Technologies for Amateur Radio. Primarily a digital voice and data mode marketed by Kenwood and ICOM.
DSW	Russian for goodbye in CW. (Do svidaniya.)
DTCS	Digital Tone Coded Squelch. Alternative to the CTCSS system.
DTMF	Dual-tone Multi Frequency. A system used to transmit and receive numeric information such as phone number, remote radio commands, etc.
Dual-band antenna	An antenna designed to be used on two amateur radio frequencies; typically, 2 meters and 70 cm.
Dualwatch	Feature on some receivers allowing them to monitor two frequencies at once.
Dummy load	A device for testing transmitting equipment without radiating a signal.
Duplex	Operation mode in which the transmit and receive frequencies are different, allowing simultaneous reception and transmission. Repeaters operate in duplex mode.
Duplexer	A device that allows duplex (bi-directional) communications on a single path. In a repeater, it isolates the receiver from the transmitted signal while allowing the receiver and transmitter to share a common antenna.
Duty cycle	The percentage of time a transmitter is operating at full power during a single transmission. Some transmission modes have inherently higher duty cycles than others.
DX	"Distance." In ham radio, usually contacts with amateurs in foreign countries, but more generally, communication over long distances.
DX Maps	Web site that shows "real-time" data on propagation by showing QSO's plotted on a map.
DX Toolbox	A suite of propagation information related software for DX'ers.
DXCC	DX Century Club. An ARRLsponsored club in which one earns membership by showing proof of contacts with at least 100 countries.
DXLab Suite	A suite of propagation information and logging software for DX'ers.

Continued on next page ...

Term	Definition
DX-pedition	A trip to a foreign, often exotic land for the purpose of operating ham radios and making as many contacts as possible, giving other hams the opportunity to "score" contacts with countries with whom ham contacts are rarely available.
Dynamic range	The difference, expressed in dB, between the noise floor of a component and the "loudest" signal it can process without excessive distortion.

E

Term	Definition
E region	The middle layer of the ionosphere (or the lowest during night hours.)
Earth ground	An electrical connection to the Earth, typically accomplished with a ground stake or, in some locales, a ground plate.
Earth station	An amateur station located on Earth, or within 50 km of Earth's surface, engaged in communications with space stations or with other Earth stations using objects in space.
Echolink	A system that connects amateur radio operators to repeaters via cell phones.
EIRP	Effective Isotropic Radiated Power. The sum of transmitter output power and antenna gain expressed in dBi, minus transmission line losses.
Electric field	A field of energy created when an electric current passes through a conductor. (See also, "magnetic field.")
Electron current	The flow of electricity from a negatively charged location to a positively charged location.
Elevation	In the context of directional antennas, a direction in the vertical plane. Most often shown on a graph of polar coordinates, with 0 degrees corresponding to the far right side of the circular graph and degrees plotted counter-clockwise around the circular graph.
ELF	Extremely Low Frequency RF. Defined by the ITU as 3 Hz to 30 Hz.
Elmer	A mentor for newly licensed (or not so newly licensed) amateur radio operators.
E-M-E	Earth-Moon-Earth propagation. Commonly known as moonbounce.
Emergency	A situation in which there is immediate danger to life, limb, or property.
Emergency traffic	Messages containing information with life or death urgency, or dealing with requests for aid in an area experiencing an emergency.
EMF	Electromotive force; the force that pushes a current through a conductor.
EMI	Electro-magnetic interference. Often caused by battery chargers and DC to AC inverters.
Emission	A transmitted signal.
End-fed	Applied to antennas with a feed point at one end, as opposed to somewhere in the middle. Since only one-half of the transmitter's output is connected to the antenna, the other half must be connected to ground or to a ground plane.

Continued on next page ...

Term	Definition
ERP	Effective Radiated Power. The sum of the transmitter's output power and the gain of the antenna, minus transmission line losses. ERP is almost always a different value than PEP, usually higher. Most ham radio legal power limits are in PEP, but the 60-meter band limit is ERP.
E-skip	Propagation by refraction from the E layer of the ionosphere.
EU	Europe.
Eyeball QSO	A face-to-face conversation with another ham.
F	
F connector	Type of connector often used in UHF installations, up to 1.2 GHz.
F region	The highest layer of the ionosphere. During the day, the F region is sub-divided into F1 and F2.
Fading	Signal reduction due to ionospheric effects.
Fan dipole	A multi-band center-fed wire antenna consisting of various lengths of wires strung between two points.
Farad	The unit of capacitance. A capacitor has 1 farad of capacitance if 1 coulomb of charge causes a potential difference of 1 volt.
FB	Fine Business. "Well done, " or "Awesome, dude."
FCC	Federal Communications Commission.
Feed point	Point where the feed line connects electrically to the antenna.
Feedline	The cable or other conductor that connects the transmitter output to the antenna.
FET	Field Effect Transistor.
Field Day	An annual event, held over the fourth full weekend of June, in which hams worldwide set up temporary stations and operate "from the field." Individuals and ham radio clubs participate. Since part of the purpose of Field Day is to showcase our hobby for the public, Field Day locations are often in public places, such as parks, museums, or beaches.
Filter	A circuit designed to pass only selected frequencies.
Fist	The sending style of a CW operator. Something like the operator's CW "accent" or "voice."
fldigi	Multiple digital HF mode software.
FM	Frequency modulation.
Foxhunting	see "bunny hunt."
Frequency	The number of times per second a signal passes through charge levels of zero, maximum positive, maximum negative, and back to 0.
Frequency Bands	(ITU) Low Frequency: 30 kHz – 300 kHz Medium Frequency: 300 kHz – 3000 kHz (3 MHz) High Frequency: 3 MHz – 30 MHz Very High Frequency: 30 MHz – 300 MHz Ultra High Frequency: 300 MHz – 3000 MHz (3 GHz) Super High Frequency: 3 GHz – 30 GHz Extremely High Frequency: 30 GHz – 300 GHz Terahertz or Tremendously High Frequency: 300 GHz – 3, 000 GHz

Continued on next page ...

Term	Definition
Frequency coordination	Local allocation of repeater input and output frequencies intended to reduce interference.
Frequency coordinator	A local group that recommends practices and implementations of frequency coordination.
Frequency Modulation	A system of transmitting information by modulating the frequency of an RF carrier wave.
Frequency privileges	The set of frequencies on which an amateur is authorized to transmit, based on their class of license.
Front-end overload	A form of distortion caused by a strong signal overpowering the receiver's "front-end", its RF amplifier.
FSCW	A CW mode in which dots and dashes are transmitted at slightly different frequencies for better intelligibility.
FSK	Frequency Shift Keying.
FSK441	See "WSJT"
FSTV	Fast Scan TV.
FT8	See "WSJT"
Fuse	A device to protect an electrical circuit from excessive amperage.
Fusion	Yaesu's proprietary digital voice and data system.
G	
Gain	Generally, a positive difference in power, such as provided by an amplifier. Gain is expressed in dB. In the case of antennas, the ratio radiated power in the direction of most radiation versus either an isotropic antenna (dBi) or a 1/2 wave dipole antenna(dBd).
Gel cell	A type of lead-acid battery that uses a gelled electrolyte, rather than liquid acid.
General coverage receiver	Usually an HF radio that will receive a wider range of frequencies than the amateur bands. Often, general coverage receivers cover frequencies from below the commercial AM broadcast band through at least the 10-meter amateur band.
GHz	Gigahertz. 1 GHz = 1, 000, 000, 000 Hz
GOTA	Get On The Air. A station available for supervised public use at events such as Field Day.
Grace period (license renewal)	The time -- two years -- the FCC allows after a license has expired during which the holder of the license may renew without retesting. The license is not valid until renewed.
Gray line	A form of HF propagation available from north to south and vice versa as the Earth passes from day to night or vice versa.
Grid square	A location identifier, such as CN88xa.
Ground plane antenna	A 1/4 wavelength vertical antenna that employs horizontal radials extending from its base.

Continued on next page ...

Term	Definition
Ground radials	Elements of a vertical antenna that are horizontal relative to the ground (or close to horizontal.) Most vertical antennas require ground radials to work properly. Radials are not safety grounding and, in fact, are not electrically connected to the Earth.
Ground rod	A copper or copper-clad steel stake driven into the ground for the purpose of making an electrical connection with Earth.
Ground strap	Heavy-duty copper strap used for ground connections, particularly in lightning ground systems.
Ground wave propagation	Propagation by radio waves that travel parallel to the surface of the Earth.
H	
HAAT	Height Above Average Terrain. For antenna and propagation calculations.
Half-duplex	See "semi-duplex"
Half-wave dipole	A basic antenna, usually consisting of two lengths of wire stretched horizontally with an insulator and the feed point in the center. The antenna is approximately one-half wavelength long relative to the desired operating frequency.
Ham	An amateur radio operator. The origin of the term is unknown, but most likely was a co-opting of a pejorative used by professional telegraphers near the time of the inception of amateur radio. Almost certainly not from "the original three amateurs" (or some variation on that), nor "the first amateur radio station call sign."
HamSphere	A virtual ham radio transceiver app for cell phones and computers. Somewhat analogous to Echolink, only for HF.
Hamvention	A gathering of amateur radio operators. Hamventions may be national, regional, or even local, and usually include vendors of amateur radio gear, speakers on various ham related topics, a swap meet, a closing dinner with a "star" speaker, and lots of camaraderie.
Harmonic	A multiple of a fundamental frequency. Colloquially, a ham's child.
Health and welfare traffic	Messages regarding the well-being of people in a disaster area. Health and welfare traffic has lower priority than "emergency" and "priority" traffic.
Heat sink	The heavy fins placed on top of a high heat component such as the final drive transistors in a transceiver. Designed to prevent excessive heat build-up.
Hellschreiber	An image transmission mode, similar to faxing.
Henry	The unit of inductance. Equal to an electromotive force of 1 volt in a closed circuit with a uniform rate of change of current of 1 ampere per second.
Hertz	See "Hz"

Continued on next page ...

Term	Definition
Hex Beam Antenna	A directional HF antenna in the shape of a horizontal hexagon. Typically multi-band, and intended for installations where low visual impact is desired.
HF	High Frequency RF. Defined by the ITU as 3 MHz to 30 MHz
HI HI	Morse code equivalent of laughing.
High pass filter	A circuit or device that allows all frequencies above a defined frequency to pass while blocking all frequencies below that defined frequency.
Homebrew	Some piece of equipment that is homemade. As a verb, the act of making equipment at home.
HT	A handheld transceiver or "handy talkie."
Hz	Hertz. The measurement of frequency. 1 Hz = 1 cycle per second.
I	
Iambic	A type of Morse code key with two paddles. Pressing one paddle sends a dash, holding it sends a series of dashes. Pressing the other sends a dot, and holding it sends a series of dots. Pressing both paddles simultaneously sends dot dash. Skilled operators can transmit very quickly with iambic keys. (The name comes from the rhythm of dot dash dot dash, etc.)
IARU	International Amateur Radio Union. The association of national ham organizations, such as the ARRL.
IC	An Integrated Circuit. A "chip."
ICOM	Amateur radio equipment manufacturer. One of the "Big Three."
IF	Intermediate Frequency. Part of the superheterodyne system of radio reception and detection.
IF shift	A circuit that shifts the IF frequency from a center frequency to help reduce interference.
IFK	Incremental Shift Keying. A digital mode.
Image frequency	A frequency offset from the desired signal by double the receiver's Intermediate Frequency. Related to image rejection.
Image rejection	A receiver specification that shows how well the receiver rejects interference from stations on frequencies that are double the receiver's Intermediate Frequency.
IMD	Inter-modulation distortion.
Impedance	The opposition to the flow of current in a circuit. Impedance consists of resistance and capacitive and inductive reactance.
Impedance matching network	See "Antenna tuner"
Inductance	The amount of magnetic storage available in an inductor. Measured in Henrys.
Inductor	A component -- typically a coil -- that stores energy in a magnetic field.
Input frequency	With regard to repeaters, the frequency the repeater is "listening to." To use the repeater you must transmit on the repeater's input frequency.
Insulator	Any material or component that blocks the flow of electrical current.

Continued on next page ...

Term	Definition
Inverter	A device that converts DC to AC.
Ionizing radiation	Electromagnetic radiation that has sufficient energy to ionize atoms, producing negative and positive ions. Ultraviolet, X-rays, and gamma rays are ionizing radiation. RF is not.
Ionosphere	A region of ionized gasses above the stratosphere.
IOTA	Islands on The Air. A ham radio activity based on operating from as many islands as possible.
IRLP	The Internet Radio Linking Project that allows amateur operators to join a global network of conversations.
ISCAT	See "WSJT"
Isotropic antenna	An imaginary "ideal" antenna that radiates perfectly equally in all directions with zero loss. Used as a reference point to compare the gain of a directional antenna. Directional gain can be specified in "dBi", meaning "dB relative to an isotropic antenna", or as "dBd", "dB relative to a 1/2 wave dipole antenna." dBd will be a lower gain number.
ISS	International Space Station.
ITU	The International Telecommunications Union. An agency of the United Nations that is responsible for issues that concern information and communication technologies.
ITU Regions	The ITU divides the world into 3 regions. Region 1 comprises Europe, Africa, the former Soviet Union, Mongolia, and the Middle East west of the Persian Gulf, including Iraq. Region 2 covers the Americas including Greenland, and some of the eastern Pacific Islands. Region 3 contains most of non-Former Soviet Union Asia east of and including Iran, and most of Oceania.
J	
JOTA	Jamboree On The Air. An annual Boy Scout amateur radio event.
J-pole	A type of omnidirectional half-wave antenna, usually for 2 meters and 70 cm, that resembles the shape of a letter J.
JT4	See "WSJT"
JT65	See "WSJT"
JT9	See "WSJT"
JTMS	See "WSJT"
Jumper	A short piece of wire or cable used to connect two parts of a circuit or two pieces of equipment.
K	
Kennelly-Heaviside Layer	Alternate name for the E layer of the ionosphere.
Kenwood	Amateur radio equipment manufacturer. One of the "Big Three."

Continued on next page ...

Term	Definition
Kerchunk (a repeater)	To press the PTT button momentarily to test whether one is hitting a repeater. A repeater that is activated will leave a tell-tale "squelch tail" that can be heard after the PTT is released. Kerchunking is bad form and illegal -- it constitutes an unidentified transmission. Be polite -- say your call sign and "testing" then release the PTT and listen for the squelch tail.
Key	Hand-operated device used to produce Morse code.
Keyer	Electronic device for sending Morse code semi-automatically.
kHz	Kilohertz. 1 kHz = 1, 000 Hz
Knife-edge diffraction	The bending of a signal by tall buildings and mountains.
L	
LCD	Liquid Crystal Display.
LED	Light Emitting Diode.
Lid	Amateur radio slang for an unskilled operator.
Lightning ground	In amateur radio, a system designed to divert lightning strikes to the antenna or tower to ground, rather than into equipment or a residence.
Li-Ion	A Lithium Ion battery.
Limiter	A circuit in a receiver that reduces variations in audio levels.
Line-of-sight propagation	Propagation in a straight line from one station to another.
Lobe	An area in the radiation pattern of an antenna.
Logbook of The World	See, "LoTW."
LOS	Loss of Signal from a satellite.
LoTW	"Logbook of The World." An online logging and QSO confirmation system run by the ARRL.
Low pass filter	A circuit or device that allows all frequencies below a defined frequency to pass while blocking all frequencies above that defined frequency.
Lowfer	An amateur radio operator who operates on the low frequency 2, 200 meter band.
LPDA	Log Periodic Dipole Array. A type of multi-band antenna.
LSB	Lower sideband modulation. In LSB, the upper sideband and the main carrier are "suppressed" - filtered out -- leaving only the lower sideband. A form of amplitude modulation.
LUF	The Lowest Useable Frequency; the lowest frequency radio signal that will reach a given destination via ionospheric propagation.
M	
Machine, The	Colloquial for a repeater.
Magnetic field	A field of energy created when an electric current passes through a conductor. (See also, "electric field.")

Continued on next page ...

Term	Definition
Maidenhead Grid System	A ham-created grid system for locating any spot on Earth within a ?grid square.? The system encodes latitude and longitude in a series of letters and numbers, such as CN88bh. The system grew out of an amateur radio meeting held at Maidenhead, England; hence, the name.
Malicious interference	Deliberate disruption of legal radio transmissions.
Maritime mobile	Amateur radio operations from aboard a marine vessel.
MARS	Military Affiliate Radio Service.
Matchbox	See "antenna tuner"
Mayday	International distress symbol on phone. From French, "m'aidez" -- "help me."
MCW	Modulated Continuous Wave. A method of sending Morse code using audio tones, rather than with momentary pulses of carrier. Commonly used by repeaters for station identification.
Meteor scatter	Propagation from the trails of ions left in the ionosphere by falling meteors.
MF	Medium Frequency RF. Defined by the ITU as 300 kHz to 3,000 kHz (3 MHz.)
MFSK	Multiple Frequency Shift Keying.
MHz	Megahertz. 1 MHz = 1,000,000 Hz
Microphone	A device to convert sound energy into electrical energy.
MMSSTV	Slow-scan TV software.
MMTTY	Multiple digital HF mode software.
Mobile	In amateur radio, generally a transceiver mounted in a vehicle.
Mode	In the context of transmission and reception, the form of modulation used to transmit information. For instance, CW, SSB, Phase Shift Keying, etc.
Modem	A "modulator/demodulator." Converts data into a radio signal and vice versa.
Modulation	The application of information to a carrier signal by systematically varying one or more dimensions of that carrier, such as amplitude, frequency, or phase angle.
Modulation Index	See "deviation."
Modulator	A circuit in a transmitter that alters a dimension (or dimensions) of a carrier wave in order to transmit information.
Monitor mode	In packet radio, a mode in which everything on a packet frequency is displayed without regard to whether the messages were addressed to the monitoring station or not.
MOSFET	Metal Oxide Semiconductor Field Effect Transistor.
MP73N	A narrow SSTV mode.
MSK144	See "WSJT"
MT63	A weak signal digital mode used in MARS net traffic.

Continued on next page ...

Term	Definition
MUF	The Maximum Useable Frequency; the highest frequency radio signal that will reach a given destination via ionospheric propagation.
Multimeter	An instrument for measuring voltage, current, and resistance. Many modern multimeters measure more values as well.
Multimode transceiver	A transceiver capable of sending and receiving multiple modes, such as CW, SSB, AM, and FM.
MultiPSK	Multiple digital HF mode software.
N	
National Electrical Code (NEC)	A set of safety guidelines for any electrical installation including towers, grounding systems, power systems, and antennas. Not all locale's local electrical code is an exact match to the NEC.
National Traffic System	An on-air network with the purpose of passing "traffic" - messages that may be between hams or between members of the public.
NB	see "Noise blanker."
NBFM	Narrow Band FM. Frequency Modulation with a deviation ratio of less than 0.5; 0.2 is common.
NCDXF	The Northern California DX Foundation, a private foundation with the purpose of supporting amateur radio and scientific projects with funding and equipment. In 1979 the NCDXF Board launched the NCDXF/IARU International Beacon Project to provide a mechanism for amateurs around the world to learn and understand more about radio propagation. The current beacon system has transmitters in 18 countries, transmitting for 10 seconds on each of five bands.
NCS	See "Net control station"
NCVEC	National Conference of Volunteer Examiner Coordinators. The national association of VEC's, and official keepers of the standard question banks for license exams.
Negative feedback	The process of feeding back a certain amount of the transmitter's output to the input of the final drive 180 degrees out of phase to prevent the amplifier from destroying itself.
Net control station	The operator in charge of conducting an on-air network.
Ni-CD	A Nickel Cadmium battery.
NI-MH	A Nickel Metal Hydride battery.
Noise	Unwanted electromagnetic energy.
Noise blanker	A circuit in a receiver that blocks intermittent noise pulses.
Noise Figure	A receiver specification denoting the difference, in decibels, between the noise output of the receiver and an "ideal" receiver with the same gain and bandwidth. Lower numbers are better.
Noise floor	The lowest level signal a receiver can "hear." Any lower level signal is below the noise floor and therefore undetectable.
Notch filter	A very narrow bandstop filter.

Continued on next page ...

Term	Definition
NPN transistor	A transistor consisting of a layer of P-type material between two layers of N-type material.
NR	Noise Reduction.
NTS	See "National Traffic System."
Null	A position in the pattern of an antenna where its reception or transmission is at minimum.
NVIS	Near Vertical Incidence Skywave. Propagation created by a high angle radiation of signal -- the closer to straight up, the better for NVIS. For short to medium distance communication via HF and, more rarely, VHF.
NXDN	Digital voice and data mode used by Kenwood and ICOM.
O	
OCF	Off-Center Fed dipole antenna. A type of multi-band dipole.
Off-center fed	Applied to antennas with a feed point that is neither at the end nor in the center. Off-center fed dipoles can be multi-band antennas.
Offset	See "repeater offset"
Ohm (Ω)	The unit of electrical resistance. 1 ohm = a resistance that will pass a current of 1 ampere when subjected to a potential difference of 1 volt. Also the unit of impedance and of reactance.
Ohmmeter	An instrument used to measure resistance.
Ohm's Law	A fundamental law of electronics that describes the proportional relationship among voltage, current, and resistance.
OM	"Old Man." Term of endearment for a male ham.
One-way communications	Communications that are not intended to be answered.
Op-amp	Short for Operational Amplifier. A high-gain solid state amplifier widely used in many applications.
Open circuit	An electrical circuit with a break in the circuit that prevents the current from completing the path to ground. An open circuit might be purposefully created with a switch, or accidentally created by a faulty connection.
Open repeater	A repeater that may be used by all hams possessing the proper license for its frequency and mode of operation.
OSCAR	Orbital Satellite Carrying Amateur Radio. Name of a series of ham radio satellites.
Oscillator	A circuit that creates a signal of a particular frequency.
Oscilloscope	An instrument used to display and measure various dimensions of a signal, most often the waveform.
Output frequency	With regard to repeaters, the frequency on which the repeater is "talking." To use the repeater, you must listen to the repeater's output frequency.
P	
P25	A digital voice mode. More formally, APCO P25.
P5	Call sign prefix of North Korea; presently the single most difficult stations with which to make a contact, although Yemen, 70, is no picnic either.

Continued on next page ...

Term	Definition
PA	Power amplifier
Packet radio	A mode of digital communication in which information is broken down into small "packets" of information for transmission, then reassembled at the receiver end.
PACTOR	Digital mode used mostly on the HF bands for text messaging.
Parallel circuit	An electrical circuit in which there are two or more paths.
Parasitic beam antenna	A directional antenna with "parasitic" (i.e., undriven) elements, such as a Yagi.
Parasitic element	A passive element of an antenna.
Part 97	The body of FCC rules and regulations that create and regulate the Amateur Service.
Passband	The range of frequencies passed by a band-pass filter.
Peak envelope power (PEP)	The average power output of a transmitter at the highest amplitude.
PEI	Peak envelope current. One half of the equation for peak envelope power.
PEP	See "peak envelope power."
Perigee	The point in an orbiting object's orbit where it is closest to the object around which is it orbiting.
Period	The time in seconds (or fractions of seconds) for a complete wave to pass a stationary point. The reciprocal of frequency.
PEV	Peak envelope voltage. One half of the equation for peak envelope power.
Phase Modulation	A system of transmitting information by modulating the phase of an RF carrier wave.
Phone	Communication by voice.
Phone emission	FCC term for voice or other sound transmission.
Phone patch	See "autopatch."
Photovoltaic	Relating to the production of electric current at the junction of two substances exposed to light. Solar cells are photovoltaic cells.
PL	See "CTCSS"
PL-259	See "UHF connector"
PM	Pulse modulation or phase modulation.
PNP transistor	A transistor consisting of a layer of N-type material between two layers of P-type material.
Polar coordinates	A graphing system that defines a point by its distance from the center of a graph (the "pole") and the angle to the point from the pole.
Polarization	The orientation, relative to the Earth, of an electromagnetic wave's electric field. A wave with both vertical and horizontal polarization is said to be circularly polarized.
Portable device	A transmitter designed to be easily operated while being carried and, for FCC purposes, with an antenna intended to be within approximately 20 cm of a human body.

Continued on next page ...

Term	Definition
Power supply	A circuit or device that supplies power for electronic equipment. Power supplies must supply adequate amperage at the correct voltage.
Priority traffic	Emergency related traffic that is not as urgent or important as Emergency traffic.
Priority watch	A feature of some receivers. The priority frequency is checked for traffic periodically, no matter where the VFO is set.
Product detector	A circuit in a receiver that detects SSB and CW signals.
Propagation	The process by which a radio wave is carried from a transmitter to a receiver. Informally, "everything that happens to our signal after it leaves our transmitting antenna."
PropLab Pro	High-end propagation calculation software.
Prosigns (Procedural signals)	Over-the-air shortcuts, such as "CQ", "73", etc. Originally used by telegraphers to speed message transmission, they are still in wide use in the amateur community.
PSK31	A type of radio-teletype using Phase Shift Keying. Very narrow bandwidth of 31 Hz.
PTT	Push-To-Talk
PWR	Power
Q	
Q	Describes the response of a resonant circuit over a specific bandwidth.
Q signals	Three-letter symbols that begin with Q and are used on CW (and sometimes on phone) to save time and increase accuracy of communication.
QCWA	Quarter-Century Wireless Association. A club for amateurs who have held a license for 25 years or longer.
QPSK	Quadrature Phase Shift Keying. A digital text and data mode.
QRA64	See "WSJT"
QRP	Low power operation, usually 5 watts or less.
QRPP	Extremely low power operation, usually less than 1 watt.
QRSS	Extremely slow-speed CW transmissions, usually less than one character per minute.
QRZ	"Who is calling me?" Also the name of a popular ham radio web site, qrz.com, which offers easy look-ups of call signs, grid squares, etc.
QSL Bureau	An organization that provides a collection and distribution point for QSL cards, for the purpose of saving everyone's postage costs.
QSL card	A paper confirmation, almost always a postcard, of a contact with another station.
QSO	A contact.
QSO Party	Ham radio contest.
QST Magazine	One of the two major general interest amateur radio magazines, the other being CQ Magazine.
Quad antenna	A type of directional antenna constructed of elements in the shape of squares.

Continued on next page ...

Term	Definition
Quad-band antenna	An antenna designed to be used on four amateur radio frequencies.
R	
RACES	The Radio Amateur Civil Emergency Service.
Radials	See "ground radials"
Radio horizon	The farthest point a VHF or higher frequency signal can travel under normal conditions. Slightly beyond the visual horizon.
Rain scatter	Propagation off a (dense) rain front.
Random wire antenna	An antenna consisting of an end-fed random length of wire, usually longer than one wavelength of the desired band of operation. An antenna tuner is almost always required for operation, as is a solid Earth ground. According to Arizona ham Patrick Lambert, W0IPL, who has studied these, an all-band (10 - 75 meters) "random" wire antenna is 74 feet long.
Receiver	A device that detects, amplifies, and retrieves the information from a signal.
Rectangular coordinates	A graphing system that defines a point by its horizontal and vertical distance from the intersection of two axes, one of which is horizontal and the other vertical. Also known as "Cartesian coordinates." Less formally, it's the "X, Y" system of graphing.
Reflected power	A product of SWR. Non-radiated power that is dissipated as heat.
Reflector element	A passive antenna element, typically located behind the driven element.
Regulator	A circuit or component that maintains a constant voltage. Part of a power supply.
Repeater	A system consisting of a receiver, a transmitter, and some means of controlling those units such that they receive signals and re-transmit them on a slightly different frequency for the purpose of extending the range of transmission.
Repeater offset	The frequency difference between a repeater's input frequency and its output frequency.
Resistance	The opposition posed by a circuit or component to the flow of electricity.
Resistor	Any material or component that opposes the flow of electrical current.
Resistive load	An electrical load that presents only resistance, with no capacitive or inductive reactance. An ideal antenna, for example, presents a purely resistive load that has the same resistance as the impedance of whatever is feeding it.
Resonant	A circuit, antenna, or even a mechanical system is said to be resonant when a relatively low amplitude, periodic stimulus of the same period as the natural vibration period of the antenna, circuit, or system produces a vibration of a large amplitude. A circuit is resonant when capacitive and inductive reactance in the circuit are both at minimum.
Resonant circuit	A circuit designed to resonate at a particular frequency, typically consisting of a capacitor, an inductor and a resistor. Multiple uses.
Rettysnitch	Companion device to the Wouff Hong.

Continued on next page ...

Term	Definition
RF	Radio Frequency. Electromagnetic waves with frequencies between 3 Hz and 3,000 GHz.
RF burn	A skin burn caused by contact with exposed RF voltages. A symptom of poor RF grounding.
RF ground	A grounding system designed to divert stray RF energy to ground.
RFI	Radio Frequency Interference. Disfunction in a piece of electronic equipment created by RF.
Rig	Amateur radio term for a transmitter, receiver, transceiver, or radio transmission system.
Ripple	In the context of a DC power supply, a waveform remaining on the DC after it is filtered.
RIT	Receiver Incremental Tuning. Allows fine tuning of the receiver frequency for better SSB reception without changing the transmit frequency.
RMS	Root Mean Square. Relates to the mathematical concept of root mean square and applies to AC voltage. Put simply, the RMS voltage is the "DC equivalent" voltage; in other words, an AC voltage of 120 volts RMS will produce the same amount of power as a DC voltage of 120 volts.
ROS Digital	Digital text mode for low signal conditions. "Ros" is the inventor's last name.
RSGB	Radio Society of Great Britain
RSQ code	Readability, Signal (strength), Quality. A signal report code similar to the RST code for amateur digital transmissions.
RST	A code of three numbers used to indicate quality of reception of a signal. R is for readability, S is for signal strength, and T is for tone and applies only to CW transmissions.
RSV code	Readability, Signal (strength), Video. A signal report code similar to the RST code for amateur television.
RTTY	Radio-teletype. A low bandwidth mode of transmitting text messages, originally to a teleprinter but now to a computer.
Rubber duck/rubber ducky	A flexible, shortened antenna used mostly on handheld transceivers. Usually covered with black plastic. Legend says they were named by a very young Caroline Kennedy when the Secret Service got the then-new antennas for their radios.
RX	Receive or receiver.
S	
S meter	A signal strength meter.
S units	Graduations on a signal strength meter.
S/N	Signal-to-noise ratio.
Safety ground	A system designed to divert high voltages on equipment cases and other items exposed to human contact directly to ground, rather than through the human.

Continued on next page...

Term	Definition
Safety interlock	A mechanically activated switch that switches off AC power to a piece of equipment when the equipment is opened.
Scan	A feature in some receivers that continually sweeps through a range of frequencies or a set of frequencies searching for signals.
SDR	Software defined radio. A form of receiver or transceiver that digitizes the incoming RF so a computer can decode the signals.
Selectivity	The ability of a receiver to discriminate between two closely spaced signals.
Semiconductor	A substance which is normally an insulator but which can transform to a conductor under certain physical and electrical circumstances.
Semi-duplex	An operation mode in which transmit and receive functions are on different frequencies alternatively.
Sensitivity	The ability of a receiver to detect weak signals.
Series circuit	An electrical circuit in which there is only one path.
Series-parallel	An electrical circuit that combines parallel and series elements.
SFI	Solar Flux Index
Shack	An amateur radio station location and associated equipment.
SHF	Super High Frequency RF. Defined by the ITU as 3 GHz to 30 GHz.
Short circuit	An electrical circuit in which a fault of some kind is taking the current to ground instead of through the desired path.
Signal reports	See "RST"
SignalLink	A brand of device for connecting a computer to an amateur transceiver via audio. The device substitutes for the computer's sound card and provides PTT control when connected with the proper cable.
Simplex	Operation mode in which the transmit and receive frequencies are the same.
SK	1. "Ceasing transmissions now, not expecting a reply." 2. A deceased amateur radio operator.
SKED	A pre-arranged contact. "I have a sked with AF7KB at 2200 Zulu."
Skip zone	An area between the end of a station's ground wave coverage and the beginning of its skywave coverage.
Skywarn	A network of trained volunteer amateur radio operators who serve as storm spotters.
Skywave propagation	Propagation via ionospheric refraction.
SLF	Super Low Frequency RF. Defined by the ITU as 30 Hz to 300 Hz.
SMA	Sub-Miniature. A type of connector, often used in VHF/UHF handheld transceivers.
Spread spectrum	A modulation and transmission system that spreads a signal over a wide bandwidth.
SOS	International distress symbol in Morse code.

Continued on next page ...

Term	Definition
SOTA	Summits on The Air. A ham radio activity based on operating from as many mountain summits as possible.
SP	Speaker
Space station	In amateur radio, an amateur station more than 50 km above Earth's surface.
Specific absorption rate	The rate at which RF energy is absorbed into a human body.
Spectrum analyzer	An instrument used to display and measure the amplitude of signals present in a particular range of frequencies.
Splatter	Interference to stations on nearby frequencies. Caused by overmodulation.
Split mode	An operating mode in which the transmit and receive frequencies are different from each other.
Split operation	Operating with transmit set on one frequency and receive on another. Often used by DX-peditions because of the enormous number of incoming transmissions they must field and answer.
Sporadic E	Semi-random propagation from anomalous regions of high free electron density in the E region.
SQL	see "squelch"
Squelch	A function that mutes audio output for certain conditions, primarily in the absence of a detectable carrier.
SS	Spread spectrum modulation.
SSB	Single sideband modulation. A form of amplitude modulation in which one sideband and part or all of the main carrier are removed from the transmitted signal.
SSN	Smoothed Sunspot Number
SSTV	Slow Scan Television. A mode of transmitting still pictures via amateur radio.
Standing wave	In physics language, the vector sum of two waves. For amateur radio, usually SWR.
Station log	A record of contacts made, signal reports, etc. Often computerized.
Straight key	One of the simplest devices for sending Morse code. See any Western movie with a scene in a telegraph office.
Sunspot cycle	The approximately 11 year cycle of average sunspot number waxing and waning.
Sunspots	Spots on the surface of the Sun that appear darker than the surrounding material. High numbers of sunspots correlate to improved HF propagation conditions.
Superheterodyne receiver	A receiver in which the incoming RF is down-converted to an intermediate frequency. The rest of the RF stages in the radio are optimized to operate at that frequency.
Susceptance	The reciprocal of reactance. Measured in siemens.

Continued on next page...

Term	Definition
SWR	Standing wave ratio. The ratio of the maximum and minimum voltages present on a transmission line. Also the ratio of any difference in impedance between the antenna and the transmission line, assuming the antenna is a purely resistive load.
SWR meter	An instrument for measuring SWR in a transmission system.
Symbol	In the context of baud rate a symbol is a signal change that indicates information. For the simplest example, consider the Morse code for the letter A, dit dah. It consists of four symbols. The dit is a symbol, the space between the dit and the dah is another symbol, the dah is a third symbol, and the (implied) space after the dah is a fourth symbol.
System Fusion	See "Fusion."
T	
Tactical call signs	"Call signs" used to identify particular functions or locations; typically during special events or emergencies. They do not replace standard call signs for station identification purposes.
Talkaround	Simplex operation.
Tank circuit	See "resonant circuit."
Telegraphy	Text-based modes such as CW, or RTTY. As opposed to phone.
Temperature inversion	A weather condition in which a layer of warm air rests on top of a layer of cooler air. Can lead to tropospheric ducting propagation.
Terminal	A piece of equipment that can replace a computer in a packet radio system.
Third-order Intercept	A receiver specification denoting how subject a receiver is to intermodulation distortion. Higher dB figures are better.
Third-party communications	Messages passed from one amateur radio operator to another on behalf of a third, unlicensed, party.
Third-party communications agreement	An agreement between the US and another country allowing amateurs to pass third-party communications between the countries.
THROB	A digital mode based on tone pairs.
Ticket	Common name for an amateur radio operator/primary station license.
Time Out Timer	See "TOT"
TNC	Terminal Node Connector modem for data communications. Also, a type of antenna connector.
TOR	Teleprinting Over Radio. There are a number of TOR modes.
Toroid	In the context of inductors, an inductor with a doughnut shaped core, whether air or some other substance.
TOT	Time-out timer. A time limit function for repeater transmitters to prevent them from overheating from long periods of uninterrupted transmission.
Transceiver	A combination unit capable of transmitting and receiving.

Continued on next page ...

Term	Definition
Transient	A brief spike of energy, especially in regards to power line supplied voltage.
Transistor	Any of a wide array of solid-state devices with multiple uses including amplification.
Transmatch	See "Antenna tuner"
Transmission line	See "feedline."
Transmitter	A device for generating RF signals.
Transverter	A device for converting received and transmitted frequencies either up or down. Might be used with a 10-meter radio to operate on 6 meters, with the received signal downconverted and the transmitted signal upconverted.
Troposphere	The lowest region of the Earth's atmosphere.
Tropospheric bending	A form of propagation typically created by mild temperature inversions.
Tropospheric ducting	A type of propagation that can occur during temperature inversions. Typically a VHF/UHF phenomenon.
TrueTTY	Multiple digital HF mode software.
TU	Thank you in CW.
Tuned circuit	See "resonant circuit."
Tuning Step	The increment that a transceiver changes frequency for one "step" of the tuning control.
TVI (Television interference)	Disruption of television reception caused by RF.
TX	Transmit or transmitter
TXCO	A Temperature Compensated Crystal Oscillator, a type of oscillator with excellent frequency stability.
B	
UFB	Ultra Fine Business. "Magnificent, splendid!" "Dude -- you crushed it."
UHF	Ultra High Frequency RF. Defined by the ITU as 300 MHz to 3, 000 MHZ (30 GHz)
UHF connector	Also known as a PL-259 plug, for coaxial cable.
ULF	Ultra Low Frequency RF. Defined by the ITU as 300 Hz to 3, 000 Hz (3 kHz.)
Unbalanced line	A feed line with one conductor at ground potential, such as coaxial cable.
Unun	A device for connecting an unbalanced source to an unbalanced load.
Uplink	Frequency used by a user to transmit to a satellite.
USB	Upper sideband modulation. In USB, the lower sideband and the main carrier are "suppressed" - filtered out -- leaving only the upper sideband. A form of amplitude modulation.
UTC	Universal Time Coordinated. While it is, for most practical purposes, the same as Greenwich Meridian Time, it is a microscopically different time standard. Most ham radio related times-of-day are given in UTC.
	Continued on next page ...

Term	Definition
V	
VA	Volt Amperes. The measure of Apparent Power (as opposed to Real Power.)
VAC	Volts of Alternating Current.
Vanity Call Sign	A call sign issued by the FCC upon a request from an operator. Certain rules and fees apply. https://www.fcc.gov/amateur-call-sign-systems
Varactor	A specialized diode who capacitance varies as the bias voltage is varied.
VCO	A Voltage Controlled Oscillator; A VCO's frequency of oscillation is controlled by an applied voltage. Typically found in phase locked loop circuits.
VE	A Volunteer Examiner; someone who administers ham radio license exams.
VEC	A Volunteer Examiner Coordinator; an organization under whose umbrella Volunteer Examiners administer ham radio license exams.
Vertical antenna	An antenna in which the radiating element is vertical relative to the ground.
VFB	Very Fine Business. "Very well done, " or "TOTALLY awesome, dude!"
VFO	Variable Frequency Oscillator. An oscillator in a receiver or transmitter that provides an adjustable frequency. Colloquially, the tuning knob.
VHF	Very High Frequency RF. Defined by the ITU as 30 MHz to 300 MHz
VIS	Vertical Interval Signaling. Information in an SSTV transmission that conveys the mode of transmission in use.
Visual horizon	The farthest point one can see by line-of-sight.
VLF	Very Low Frequency RF. Defined by the ITU as 3 kHz to 30 kHz.
Volt (V)	The unit of electromotive force. 1 volt = the difference of electric potential that would drive 1 ampere of current through 1 ohm of resistance.
Voltage	Electromotive force. The amount of charge in one location relative to another location. Informally, "electrical pressure." Measured in volts.
Voltmeter	An instrument used to measure voltage.
Volunteer Monitor	A member of the FCC Auxiliary. A volunteer amateur operator who monitors the airwaves for FCC rules violations.
VOX	Voice Operated Transmission. Allows the presence of sound at a microphone to operate the PTT.
VSWR	See "SWR."
VUCC	"VHF/UHF Century Club." An ARRL award for verified contacts with a minimum number of Maidenhead grid locators per band.
W	
WAC	Worked All Continents. An award issued by the IARU to those who prove they have had a contact with at least one ham on each continent.

Continued on next page ...

Term	Definition
WAN	1) Wide Area Network. "HAMWAN" is a form of wide area "wi-fi." 2) Worked All Neighbors. A station that generates many complaints about interference with neighborhood telephones and televisions. (Don't get this award.)
WARC	World Administrative Radio Conference.
WARC bands	Allocated to amateurs by the WARC in 1979, they are the 30 meter, 17 meter, and 12-meter bands.
WAS	Worked All States. An ARRL award to those who prove they have made a contact with at least one station in each U.S. State.
Watt (W)	The unit of measurement of the use of electrical power. 1 watt = a current of 1 amp at a voltage of 1 volt.
Wattmeter	An instrument for measuring power; in the case of amateur radio, this is usually the output of the transmitter. Wattmeters can be directional, measuring either forward or reflected power.
Waveguide	A carrier of microwaves to and from a radio to an antenna. A waveguide is a structure that guides electromagnetic waves (or, in other applications, sound waves) with little loss of energy by restricting the expansion of that wave. Also, a form of propagation of radio signals in the 3 to 30 kHz frequency range.
Wavelength	The physical length of one cycle of a signal. In any given medium, all radio waves travel at the same speed, so lower frequencies have longer wavelengths. To determine the approximate wavelength in meters of any frequency, divide the frequency, expressed in MHz, into 300.
WAZ	An award from CQ Magazine recognizing amateurs who have proof of contact with a particular number of the "CQ DX Zones." The number varies by band and mode. See http://www.cq-amateur-radio.com/cq_awards/cq_waz_awards/cq_waz_award_types.html
WFM	Wideband Frequency Modulation. A system of sending information by varying the frequency of the carrier by a greater amount than Narrow-band Frequency Modulation, i.e., a deviation ratio greater than 0.5. The maximum deviation ratio of WFM for amateur radio is 1.0; in other words, the carrier may deviate no farther from the center frequency than the frequency of the modulating signal. If sending a 1000 Hz tone, the carrier may deviate no more than 1000 Hz.
Wires-X	"Wide-coverage Internet Repeater Enhancement System." An internet linking feature of the Yaesu System Fusion digital voice and data system.
Wouff Hong	A torture device of unspeakable horror used for punishing hams who fail to follow common courtesy on the air. Many ham conventions feature a secretive late-night ceremony in which they are inducted into the mysterious Royal Order of the Wouff Hong.

Continued on next page ...

Term	Definition
WPX	"Worked Prefixes." An award from CQ Magazine recognizing amateurs who have proof of contact with a particular number of stations whose call signs begin with the many international call sign prefixes around the world.
WSJT	The Weak Signal Joe Taylor suite of weak signal digital transmission modes. The complete system consists of WSJT and WSJT-X. Various modes can be used for meteor scatter, moonbounce, and other weak signal situations, including general purpose communication in poor propagation conditions. The software suite is available free at https://physics.princeton.edu/pulsar/k1jt/wsjtx.html
WSPR	Weak Signal Propagation Reporter. Part of the WSJT suite.
WSPRNet	A reporting tool for the WSPR system.
WWV	A radio station located in Fort Collins, CO, and operated by the National Bureau of Standards. It broadcasts the time of day, propagation reports and other information, and serves as a frequency standard.
WWVH	A station similar to WWV, broadcasting from Hawaii.
WX	Weather.
X	
XCVR	Transceiver.
XIT	Transmitter Incremental Tuning. A feature on some HF radios that allows fine tuning of the transmit frequency without changing the receive frequency.
XTAL	Crystal.
XYL	An "ex-YL." Vintage ham term for "wife." "YL" is far more commonly used now, for reasons that are probably obvious.
Y	
Yaesu	Amateur radio equipment manufacturer. One of the "Big Three." Usually pronounced YAY-zoo.
Yagi antenna	A type of directional antenna with a single dipole radiating element, a reflector, and at least one director element, all mounted parallel to each other.
YL	"Young lady." Might describe any female, but especially a female ham, or a ham's wife.
YF	Short for "wife."
Z	
Zed	The letter "Z."
Zener diode	A specialized diode used for power supply regulation.
Zero beat	To tune precisely to a received frequency.
Zulu	In the context of time-of-day, UTC

Table 12.1: Amateur Radio Glossary

Index

Fast Track Ham Radio Facts, 8, 174
Grounding and Bonding for the Radio Amateur, 91
National Electrical Code, 91
Standards and Guidelines for Communication Sites, 91
12-volt convenience outlet, 64

AF gain, 165
AGC, 168
Air Boss antenna launcher, 148
airbags, 46
Alinco DX-SR8T, 157
Alpha Antennas, 153
Amateur Radio Glossary, 191
Amazon for ham radio gear, 51
antenna analyzer, 72
antenna placement, mobile installation, 66
antenna rotor, 82
antenna site, fixed station, 82
antenna tuner, 161
ARRL, 4, 23, 41, 91, 118, 131, 140, 174, 177, 193, 198, 203, 205, 217, 218
ARRL Repeater Directory, 23
attic antenna, 152

audio processing, 167
automatic gain control, 168

band switch, 165
band-pass filter, 167
Baofeng accessories, 27
Baofeng Tech, 50
Baofeng UV-50X2, 43
Baofeng UV-5R, 27
Baofeng, programming, 28
Bernstein, David, AA6YQ, 179
bonding, 97, 106
BridgeCom Systems, 135
BTech UV-50X2, 43
BTech UV-5R, 27
Buddipole, 39
building HF station, 142
butt splice, 57
BuyTwoWayRadios.com, 50

C4FM, 132
cable ties, 58
CAR control, 168
Casler, Dave, 186
center-tap power supply, 160
CHIRP, 24, 32, 195
coaxial cable, 85
coaxial cable loss, 87

colinear antenna, 26
Collins KWM-2, 157
Comet SBB-5NMO, 52
common mode current, 102
common mode currents, 101
computer time, FT8 and, 187
contest exchange, 130
contest rules, 128
contest strategies, 128
contesting, HF, 176
contesting, VHF/UHF, 128
continuity checker, 71
cookie sheet, 27
counterpoise, 149
crimp connector, 58
critical frequency, 180
CSCE, 3
CTCSS tone, 22
cupholder mount, 70
CW, 171, 180

D-Star, 132
DCS tone, 22
detached control head, 42
digital hot spots, 134
digital modes, HF, 183
digital voice modes, 131
digital voice repeater linking, 131
direct buriable, 155
Distracted driving laws and ham radio, 41
DMR, 132
downlink tone, 24
Drive control, 165
dual-band handheld transceivers, 25
Dual-band mobile station, 41
DX, 177

DX Engineering, 50, 155
DXLab Suite, 179
DXView, 179
DXView World Map, 179

EchoLink®, 121
ECHOTEST, 122
EI2KC, 179
Elma, 7
Elmer, 7
external antenna (for HT), 26

FCC, 6
FCC (Baofeng and), 28
FCC Part 97, 7
feed line, fixed installation, 84
Fender-style Antenna Mount, 53
ferrite choke, 102, 103
flagpole antenna, 150
flat-topper antenna, 144
frequency read-out, 165
FT8, 186
FT8 screenshot, 187
fuses, 50, 65

G5RV antenna, 146
gable mount, 83
GigaParts, 50, 155
gooseneck mount, 70
grommet, 64
ground loop, 110
ground rod, spacing of, 96
ground rods, 96
ground rods, driving, 96
ground, equivalent circuit of, 94
ground, HF station, 91
grounding bus, 100

grounding, lightning protection, 104
grounding, RF, 99
grounding, safety, 97
grounding, VHF/UHF station, 91

HAAT, 82
Ham Radio Concepts, 63
Ham Radio Deluxe, 155, 185
Ham Radio Outlet, 50, 155
Hamstick, 152
Heater control, 165
heliax, 88
Hertz, Heinrich, 6
HF antennas, 144
HF definition, 139
HF feed line, 154
HF power supply, 160
HF propagation, 140
HF switch-mode power supply, 161
HF transceiver controls, 163
Home Depot, 26
homeowners' association, 42
horizontal loop antenna, 146
HyGain DX-88 antenna, 149

ICOM, 42
ICOM IC-718, 158
ICOM IC-7300, 159
ICOM IC-7851, 156
IF shift, 165
inverted V antenna, 145
ionosphere, 140
Islands on the Air, 42

J-pole Antenna, 38

K9JEB, 38

KE0OG, 186
Kenwood TH-D72A, 35
Kenwood TM-D710GA, 42
Kenwood TS-520S, 159
KJ4YZI, 63

ladder line, 38
lightning, 93
lightning/surge protector, 89
lip mount installation, 67
listening on the input, 113
Load control, 165
Logbook of the World, 178
long-wire antennas, 144
long-wire antennas, erecting, 147
LotW, 178

magnetic loop antenna, 153
Marconi, Guglielmo, 6
meter (on transceiver), 167
MFJ, 155, 161
microphone gain, 167
mobile antenna grounding, 47
mobile antenna placement, 47
mobile antenna, selecting, 51
mobile feed line, 54
mobile magnetic antenna, 26
mobile power connectors, 57
mobile radio installation, 45
mobile radio, powering, 56
Mode switch, 167
Mototrbo, 132
mounting with velcro or double-sided tape, 46
mounts, Lido, 70
MUF, 183
multi-mode VHF/UHF transceiver, 77

net control station, 114
net control transcript, 118
nets, HF, 174
nets, traffic-handling, 118
nets, VHF/UHF, 118
NMO connector, 52
noise blanker, 168
non-penetrating roof mount, 84

ohmmeter, 71
operating on HF, 171
order of operations, mobile installation, 48

permeability, 103
phonetic alphabet, 115
PL-259 connector, 52
PL-259 connectors, installing, 68
Plate control, 165
Pofung UV-5R, 27
power hammer, 97
power supply, 78
power supply, HF, 80
power switch, 164
Powerpoles©, 80
preppers, 4
professional installation, mobile radio, 47
programming cable, Baofeng, 34
programming software, Baofeng, 32
propagation prediction, 180
propagation software, 179
PropView (Propagation View), 179
PSK31, 184, 185
PSK31 screenshot, 186

QSL cards, 178
QYT KT-8900D, 45

QYT KT-980, 43

R-CTS, 24
radial plate, 150
radial system, 149
rain gutter antenna, 152
remote operation, 154
repeater listing, 23
repeaterbook.com, 23
repeaters, 21
repeaters, finding, 22
repeaters, working, 114
reverse, 113
RF attenuation, 167
RF gain control, 169
RF grounding, 99
RIT, 165
RST Signal Reporting System, 173, 174
RT Systems, 34
rubber duckie, 26, 37

safety grounding, 99
screwdriver antenna, 152
setting up for digital, HF, 183
short circuit, feed line, 71
SignalLink, 184
Silver, H. Ward, 91
simplex frequencies, 125
simplex operation, 125
Slim-Jim Antenna, 38
Slinky antenna, 152
sloper antenna, 145
Smoothed Sunspot Number, 180
Software Defined Radio, 159
soldering iron, 69
soldering mobile connections, 58

SOTA, 127
space weather, 180
split wire loom, 58, 65
squelch, 113
SSB phone, 172
Standby switch, 168
stealth antennas, 42
Summits on the Air, 42
Survival Antenna, 38
SWR, 71, 72, 146
SWR Meter/Wattmeter, 73
System Fusion, 132

talk group, 131
talk groups, 131
tape measure Yagi, 128
Tigertronics, 184
time.is, 187
tone (repeater), 22
ToneSql, 24
tools for mobile installation, 58
tower, 84
Tram dual-band antenna, 26
Trunk Lip Mount, 54

tuning mobile antenna, 74

under-seat mobile mounting, 63
upgraded antennas for HT, 37

vertical antennas, 149
VFO, 113, 165
VOX, 168
VOX delay, 168

waterproofing tape, 155
wavelength, 102
WebSDR.org, 172
window feed-through, 155
Window Gap Jumper, 53
window line, 38
window pass-through, 51
Wouxun KG-UV920P, 43

Yaesu FT-2DR, 35
Yaesu FT-65R, 35
Yaesu FTM-400DR, 41
Yaesu System Fusion, 132

Zerofive vertical antenna, 150

About the Authors

Michael Burnette, AF7KB, started playing with radios at an early age when he found the plans for a "Foxhole Radio" in a comic book and built one. It used a toilet paper tube for a coil form and a rusty razor blade and safety pin as a detector. It almost worked.

He would go on to become a commercial broadcaster, then an adult educator. As a seminar leader he led classes in leadership and communication all over the world, primarily in Hong Kong and China.

Kerry Burnette, KC7YL, came later in life to the world of radio, but embraced it enthusiastically and serves as net control for our local club's weekly net as well as leading the club's weekly YL net.

Kerry was the original inspiration for the first *Fast Track* license course, which started as a set of notes for her to use preparing for her Technician exam.

A graduate of the University of Washington, Kerry spent a decade teaching high school math and science and holds a Masters of Education degree.

Michael and Kerry often appear at major hamfests, where Michael is a sought-after speaker.

Together, they have created

The Fast Track to Getting Started in Ham Radio
The Fast Track to Your Technician Class Ham Radio License
The Fast Track to Mastering Technician Class Ham Radio Math
The Fast Track to Your General Class Ham Radio License
The Fast Track to Mastering General Class Ham Radio Math
The Fast Track to Your Extra Class Ham Radio License
The Fast Track to Mastering Extra Class Ham Radio Math
The Fast Track to Understanding Ham Radio Propagation
The Fast Track Ham Radio Facts Book

Most *Fast Track* programs are available in paperback, e-book, and audio formats from major online retailers.

www.ingramcontent.com/pod-product-compliance
Lightning Source LLC
Chambersburg PA
CBHW071355210526
45465CB00001B/96